Praise for Sound Healing

"What a joy to find such practical treatment protocols so well presented in both book form and shown on the DVD! As a Doctor of Traditional Chinese Medicine, I often incorporate these acupoints into the sound healing treatments I give. The sound vibration of the Ohm Tuning Fork is unifying and brings harmony to both heart and mind.

Marjorie's Manual & DVD are fantastic educational tools for both the health care professional & anyone intent on learning to use sound for personal benefit. The Ohm frequency will transport you to a familiar place – that of being enveloped & held by the Great Mother herself! A lovely introduction to the nature of vibrational healing."

~ Samantha Jennings, Dr.TCM, R.Ac.
Practitioner & Educator of Oriental Medicine &
Ohm Therapeutics Sound Healing Instructor

"Having used the Ohm Tuning Forks for several years, they have become an indispensable part of my healing tool kit. Many patients proclaim how much better they feel right away and that it is the favorite part of their treatment. The tuning forks work, patients love the treatment, and it resonates with people on a very deep and profound level to initiate healing."

~ Jimmie McClure, DC
Holistic Chiropractor & Meridian Therapy Practitioner

"While music in the key of Ohm plays, the application of the Ohm Tuning Forks literally attune the physical, environmental and spiritual elements of one's being. The message is from the Earth; the body feels whole, relaxed and balanced. As a homeopath I understand the function of resonant properties in healing.

During treatment, the experience is totally fluid and grounded at the same time. In the days and weeks following, the feeling is one of wholeness and communication with the vibratory earth elements, and there is a definite awareness that the healing channels of the body have been awakened."

~ Donna Caulton, MSN, FNP
Homeopath, Nurse and Visual Artist

OHM TUNING FORKS BEAUTIFULLY EFFECTIVE.
"*Integrating the use of tuning forks into your massage session can help your clients reach a deep state of relaxation and experience a restored sense of well-being. I used two Mid Ohm Tuning Forks symmetrically on acupressure points with the intention to move stagnant energy. I also tried them for relieving joint pain in combination with de Muynck's CD,* There's No Place like Ohm, *and found them beautifully effective.*"

~ Massage Therapy Journal, Spring 2008

"*Sound Healing is the perfect musical mojo for a world that is out of sync with the melody of the earth. Not just for the professional practitioner, this Manual and DVD are as indispensable as the Ohm tools described; it is a gift to a world much in need of re-tuning with earth-song. A sound innovator whose training spans the gap between tribal and classical methods, de Muynck adds to the ancient instruments traditionally used to maintain the important synchrony of the human body and the earth.*"

~ Karen Guerin
Polarity, Sound Healing, and Cranial-Sacral Practitioner, Poet and Educator

Sound Healing
Vibrational Healing with Ohm Tuning Forks

Works of Related Interest

CDs
There's No Place Like Ohm (Lemniscate Music, 2002, Remix w/extended play, 2008)
There's No Place Like Ohm Vol. 2 (Lemniscate Music, 2005)
In the Key of Earth (Sounds True, 2007)
Vibrational Healing Music (Sounds True, 2009)

Books
Acutonics®: There's No Place Like Ohm, Sound Healing, Oriental Medicine and the Cosmic Mysteries (Devachan Press, 2003) by Carey and de Muynck

Future DVD Releases
Onsite Chair Massage
Self-Treatment
Sound Facial™

Please check our website and sign up for our newsletter to be informed of future releases:
www.soundhealingtools.com

Sound Healing
Vibrational Healing with Ohm Tuning Forks

A Practical Application Manual and DVD

Marjorie de Muynck, M. Mus., LMP
Illustrated by Linda Marie Waller

Santa Fe, New Mexico

Published by LEMNISCATE MUSIC
Copyright © 2008 Marjorie de Muynck
Illustrations © 2008 Linda Marie Waller
Cover Design by Janice St. Marie
Book Design by Linda Marie Waller
Cover Photograph by Wendy McEahern Photography
Back Cover Photograph by Carolyn Hall Young
Color Section Photographs by Lennart Nilsson/Albert Bonniers Förlag AB, Forest Woodward, Manuela Millar mapped from NASA earth image, and Wendy McEahern.
Stylized treatment photographs by Solar Law.
Sound healing tool photos by Wendy McEahern.
Bewitched photograph courtesy Sony Pictures Television.
Cadence and *Wintersong* paintings courtesy of Donna Caulton.

All rights reserved. This Manual and DVD, or parts thereof, may not be reproduced in any form without written permission.

First Lemniscate Music trade paperback edition: November 2008
Lemniscate Music trade paperback ISBN 978-0-615-23943-9

Printed in the USA with soy-based ink on recycled certified green materials.
ECO DISC DVD is made of 33% less plastic and chemicals. Using new manufacturing processes, which use 50% less petroleum, the Magtek ECO DISC™ reduces the need for petroleum based raw materials. Manufactured in the USA.

 Information and products are not intended to diagnose or treat. This Manual and DVD are not intended to replace the services provided by licensed health care providers. For medical conditions, please consult your physician.

To learn more about our sound healing tools and educational workshops contact:

SOUND UNIVERSE, LLC
Lemniscate Music, Publishing
Ohm Therapeutics, Education
1A Plant Farm Road
Santa Fe, NM 87506
info@soundhealingtools.com
www.soundhealingtools.com

For the regenerative healing of the Earth and all her inhabitants

Acknowledgments

This Manual and DVD would never have been started or completed without the consistent support and hard work of my publisher and editor Linda Marie Waller. She believes in this work, and has been witness to my healing with sound as well as the hundreds of testimonies she has received at Lemniscate Music and Sound Universe over the years. She is a lovely human being.

I gratefully acknowledge all of the people and animals I have treated over the past 26 years. Without these invaluable treatments I would have only speculation and conjecture.

Special thanks to Samantha Jennings, Dr. TCM, R.Ac. for her consistent and core support of this medicine. Samantha is a tremendously knowledgeable Instructor of this system. Thank you to my dear friend Donna Caulton MSN, FNP, an extraordinary homeopath, nurse and visual artist, who has assisted me through a healing miracle, and who inspires my work through her resonance paintings. And for her brilliant observations of life, I thank Liz Durkin, MAc, LAc, LMP. Her loving support and extremely helpful Vedic readings have been my compass and guided me through many difficult years. Thank you to Dr. Jim McClure, Holistic Chiropractor and Meridian Therapy Practitioner, for our shared sound research. Jim Cameron, heartfelt thanks for his ongoing support and guidance. Annie Lindberg and Paul Dwyer, whose steadfast and generous support are deeply appreciated. Special thanks to Janice St. Marie for her beautiful graphic design skills and Wendy McEahern, many thanks for her excellent photography skills. Thanks to Hall and Young productions. And many thanks to Lorin Parrish, Director of the New Mexico Academy of Healing Arts, for supporting this work and offering our Sound Healing classes through the academy's continuing education program.

And a very special thanks to those beings and spirits who have taught me and helped me awaken to sound, vibration, music and healing; both those who are here with us and those who have gone before us:

Hans Cousto for his important work with elliptical orbits; Makira Enrico, who helped reawaken my interest in tuning forks; Kay Gardner blessed carrier of the story of sound; Joseph Rael, "ah-eh-eee-oh-uu;" Dr. Elizabeth Kubler-Ross who inspired and encouraged me to keep playing music; Hans Kayser, patron saint of world harmony; Ellen Fullman and her amazing Long Stringed Instrument; John Coltrane who heard Ohm and told his musical story; Rachel Carson, environmental pioneer and very brave soul; Brant Secunda for whom I am forever grateful for his Shamanic healing; and Dr. Helen Caldicott, a warrior for the health of the Earth.

Without the amazing and regenerative healing power of the Earth – and the innate healing intelligence of our bodies, this work would not be possible.

Foreword

The sound vibration of Ohm has long held a revered place in humanity's cultural and religious practices. A sacred mantra, the syllable of Ohm is used to create a peaceful atmosphere for prayer and intention to be expressed, and is thoroughly ensconced in contemporary healing practices. Ohm is often chanted at the beginning and end of a yoga class to quiet the mind and promote a sense of unity, and to align teacher and student through shared resonance with the same frequency. Marjorie de Muynck has developed a vibrational healing system that features this historic tone. In this system, the Ohm tone helps set the intent of practitioner and patient by aligning them with this therapeutic sound healing frequency. For reasons that will become clear as you read on, Ohm is truly the quintessential earth tone, and as Marjorie describes, our musical center of gravity.

Marjorie de Muynck has been devoted to music and sound healing for over 30 years. I met Marjorie in 1999 during my initial year of acupuncture training at the Northwest Institute of Acupuncture and Oriental Medicine. I was immediately struck by her passion for alternative medicine and her knowledge of many healing traditions and modalities. During this time Marjorie was diagnosed with breast cancer. Her willingness to openly discuss the impact of her diagnosis and share her own process inspired many students and colleagues. She became an example of how to practice the medicine that she was teaching. We learned about the archetypal wounded healer within all of us.

Like a shamanic initiation, the archetype of the wounded healer is one in which a healer is faced with a mysterious illness that is often difficult to heal. It is the process of going beyond the illness, through the darkness and suffering that ultimately transforms the healer and reveals new paths of healing for both self and others. It is as if the wound itself drives one to the inner journey that becomes the transformation.

Over the years Marjorie and I have shared our processes and discoveries which seem to be leading us along the same winding and unpredictable path. So we were not surprised when last year our journeys took a similar turn. We were both diagnosed nearly simultaneously with chronic Lyme disease. Unfortunately, the allopathic medical model has not had much to offer and does not yet recognize or understand the complex pathology of this disease. Marjorie and I continue to discuss our experiences with alterative treatments and share the results. What strengthens each of us is listening to our intuition, using vibrational medicine such as sound and homeopathy, and staying grounded in nature and connected to the Earth.

As Marjorie was finishing her CD, *In the Key of Earth*, I marveled that she had the energy to complete not only her graduate studies at Boston University, but also continue her recordings of Ohm in the natural world while healing her own body. One might say she was driven by a sound vision emanating from the Earth itself and the energy of Ohm. We are fortunate that Marjorie is attuned to the presence of this sound and sensitive to its vibration in nature. She provides an intriguing theoretical basis for incorporating this sound vibration into contemporary healing practices. Most healing systems don't address the inter-relationship between an individual's health and the health of their environment. I am very impressed that the *Ohm Therapeutics* system focuses on how we are influenced by the cycles and rhythms of the Earth.

This book and DVD eloquently bridge ancient knowledge of Ohm with scientific understanding to present a refined healing system that is simple yet comprehensive. It is a manual for beginners and seasoned healers alike. As an acupuncturist I often use tuning forks instead of needles because they direct energy in a gentle and self-regulating way. Marjorie acknowledges the body's innate healing intelligence, and this theme underlies her healing perspective. The beauty of this system comes at a much needed time in our personal and planetary evolution. It is refreshing to read a work that examines the benefits of a musical system based on the historically rich tone of Ohm. During this time in our evolution, when we are beginning to understand just how far we have strayed from nature and her cycles, Ohm is exactly what is needed. It points the way home.

As an acupuncturist and body worker, I find the combination of music and tuning forks tuned to Ohm to be uniquely powerful. The whole room becomes charged, and both the patient and practitioner receive the calming and energizing benefits of Ohm. Like her predecessors in the Hindu and Buddhist traditions, for example, where the mystical symbol of Ohm has long been revered, its meaning explored in sacred religious texts and prayer, and whose sound continues to be chanted throughout the world, Marjorie provides us with another way to experience Ohm. Through her musical compositions and the use of tuning forks, we can explore Ohm at home and in our practices. Marjorie's work will inspire those who would like to incorporate music and sound into their lives. Practitioners from all walks will find her treatment protocols easy to implement. Like the ancients before her, she leaves us a map so that we may understand our connection to this familiar earth tone.

I find it very interesting to note that Marjorie's astrological chart reveals something of a cosmic destiny at work in her sensitivity to hearing Ohm and pioneering this therapy. What stands out to me as a Vedic astrologer is that the planet Mars, the indicator of an individual's energy and drive, falls in its exaltation sign of Capricorn and is placed in the 8th house of transformation.

In music an octave is an interval of 8 diatonic degrees between two like tones, the higher of which has twice as many vibrations per second as the lower. An octave marks the place of transformation between musical intervals, like an evolutionary step up or down in frequency. According to astrological traditions, the 8th house represents the field of activity relating to hidden things, death/longevity, and transformation – the evolutionary steps of our soul. When turned on its side the number 8 becomes the symbol for infinity, esoterically connecting 8 with the process or steps of evolution. Thus 8 is the number for change. For example in the Taoist tradition, the *bagua* contains eight trigrams that represent how energy shifts from yin to yang and back again.

Marjorie's drive to understand the hidden and subtle effects of sound and vibration, as well as their transforming power, is in part triggered by the 8th house Mars. It is further intensified by a conjunction of the karmic north node, *Rahu*, which has the power to amplify. Rahu symbolizes our destiny in the material world, and its placement in the house of the hidden things reveals Marjorie's shamanic ability to access information from these etheric realms and manifest them in the material world. This intense planetary line up of Mars and Rahu is heavily influenced by a group of stars in Capricorn that the ancient seers of India called *Shravana*. According to Vedic Astrologer Dennis Harness:

> *Shravana is derived from the Sanskrit verb sru which means "to hear" and is symbolized by the ear. It reflects the ability to hear the subtle etheric realms. The cosmic sounds of Krishna's flute, a bell, or the cosmic OM may be audible to the Shravana native. It represents communication of knowledge that helps transcend the material world.*
> —Dennis Harness, *The Nakshatras: The Lunar Mansions of Vedic Astrology*, (Lotus Press, 1999).

This description certainly reflects Marjorie's magic and life path. It is said that people with planets in Shravana possess *samhanana shakti*, or the power of connection. Marjorie de Muynck has used this stellar gift of 'listening' to bring forth a new understanding of the transformative power of Ohm. Her work teaches us how to reconnect to both the subtle etheric effect of Ohm as well as its vibrational manifestation on the physical plane. *Ohm Therapeutics* is a transformative healing system that teaches us to listen and connect with the most elemental energy of the Earth and Cosmos: Ohm.

—**Elizabeth Durkin, MAc, LAc, LMP**
Seattle, Washington, August 2008

Introduction

I received my first tuning fork as a child. In the early 1960's, my father gave me the bright shiny object to tune my guitar. After tuning my guitar I began to place the tuning fork on my head, chest, legs – I was fascinated by how the vibration felt on my body. I have to admit, it was one of those moments when I felt as if I was doing something strange or amazing that no one would understand and would be considered weird, so I kept it to myself. I was 11 after all. When watching the TV show *Bewitched* one evening, Dr. Bombay began using a tuning fork on Samantha to cure a malady! It was truly one of those unforgettable and strangely transforming moments. I felt ecstatic, validated and curious.

Later, in 1980, the piano player in my band surprised me by pulling a tuning fork out of her pocket to treat a migraine headache I had been suffering with for days. The applied sound vibration of the tuning fork helped break the pain cycle. I was just beginning to study alternative medicine at the time and this experience deeply influenced my direction as I explored the relationship between music and the healing arts.

In 1995, at the Northwest Institute of Acupuncture and Oriental Medicine in Seattle where I was teaching Shiatsu, Hara Diagnosis and Kundalini Yoga, I was involved in a creative partnership with the clinical dean Donna Carey. We started the *Acutonics*® Sound Healing system using tuning forks on acupuncture points. Originally the tuning forks were used to treat needle sensitive patients in the school clinic but it became popular very quickly with those coming through this large public clinic and the dozen other satellite clinics in the greater Seattle area. There seemed to be a thirst for this type of medicine. We started teaching classes for acupuncturists and eventually began teaching massage therapists, nurses, chiropractors, energy workers, and other healthcare practitioners.

Three years ago I lost most of my T6 vertebra when my spine was damaged from cancer. The use of sound vibration on the area has helped relieve pain and increase the bone density of the remaining vertebrae, strengthening my overall spine. As a result, I have been able to return to most of my previous physical activity. I remember years ago leaving a venue with loud music. The bass and low tones followed me out, bending around walls and even buildings; all the while I could feel the sound waves traveling through my body. I recall thinking at the time that low frequency might be an effective sound healing tone for the body's joints and vertebrae. Recent research validates that low frequency sound vibration not only heals bone and alleviates pain but it actually increases bone density.

During my journey with breast and spinal cancer, sound continues to be an integral part of my healing. My daily treatments help provide physical and spiritual balance. Serious illness can leave you feeling disconnected and ungrounded. After receiving Sound Healing treatments, I feel energetically rooted and connected to a larger healing force.

Research shows that cancer cells are weak and floppy. The Ohm vibration strengthens my center and helps to resolve the disharmony of disease. Follow up scans continue to show strong and healthy cell activity – free of cancer.

During this time of great change on our planet, I believe it is more important than ever to live in harmony with the Earth. By connecting with the elements and observing Nature's glorious rhythms and cycles, we can experience profound physical and spiritual healing. While we long for a panacea, an elixir to cure our ailments, I am convinced that we will progress much further when, as a society, we address the ills that affect our environment. When we think of how our immune systems are challenged, we must consider that the Earth's immune system, including many of her beautiful species, has also become challenged.

A successful practitioner is a facilitator who acknowledges and nurtures the inherent healing ability of the person being treated. Rather than imposing a belief system or modality, the practitioner creates a sacred space and sound environment. From within this space, listening enables the body to find its healing way. In this manner, we become facilitators—conduits for the Universal healing intelligence that surrounds us all.

For those of you drawn to this work, there is great reward and responsibility not only in how you treat yourselves and your clients, but in your relationship to the Earth, and all its inhabitants.

—**Marjorie de Muynck, M. Mus., LMP**
Santa Fe, New Mexico 2008

Vibrational Healing with Tuning Forks™
ohm therapeutics

Ohm Therapeutics is a basic yet comprehensive sound healing system featuring the universally recognized vibrational frequency of Ohm and its overtones. Based in science and spiritually upheld, Ohm is an ancient and sacred tone and mantra celebrated by cultures throughout the world for its healing properties. To resonate with Ohm is to unite with the life-supporting energy of the Earth, which positively affects our biological rhythms and circadian clock. Through sympathetic resonance, we begin to sync and entrain with natural cycles. Aligning with these earth rhythms enables us to find balance.

The application of sound vibration to the physical and subtle body opens the energetic pathways where the Qi or natural life force flows. As a result, energy blocks are removed—increasing the flow of Qi—facilitating homeostasis through which profound healing begins. Ohm Therapeutics features **Music plus Tuning Forks** tuned to this sacred healing frequency.

SANTA FE, NEW MEXICO

Contents

Acknowledgements ix
Foreword xi
Introduction xv
Ohm Therapeutics xvii
Guiding Principles xxiii

Part I
Sound Healing
Benefits 3
Our Vibratory System 3
Our Earthly Body 3
Intuition and Intention 4
Resonance and Entrainment 5
Sympathetic Resonance 5

Part II
Sound & Vibration
Vibration and Form 9
Electromagnetic Frequencies 11
Threshold of Noise 12
Medical Application of Frequency 12

Part III
Ohm, Octaves & Overtones
Ohm and C 16
Ohm ~ Our Musical Center of Gravity 17
Mid Ohm Frequency ~ 136.1 hz 18
Keynote of Ohm 18
Contemporary Ohm Tuning ~ Singing Bowls 19
Resonance with Ohm 19
Octaves 20
Overtones 21

Part IV
Tuning Fork Technique
Acoustic Advantage 23
Parts of a Tuning Fork 24
How to Hold a Tuning Fork 24

How to Activate a Tuning Fork 25
Wearing a Practitioner Activator 25
How long will a Tuning Fork Vibrate? 25
Number of Applications 26

Mid, Low & Osteo Ohm Tuning Forks
Application Process: Find, Palpate, Eye, Activate, Place 26
Applying to the Body: Mid, Low & Osteo Ohm Tuning Forks 26

High Frequency Ohm Tuning Forks
Application Process: Activate, Direct 27
Applying High Frequency Ohm Tuning Forks to Energetic Body 27

Rhythm and the Sound Healer 28
A Musical Rest 29
How to Care for Tuning Forks 29
Ohm Therapeutics Tuning Forks 29
A Quick Look at Tuning Forks 31

Part V
Essential Tools
Basic Practitioner Tools 33
Highly Recommended Tools 33
Two Mid Ohm Tuning Forks 33
Low Ohm Tuning Fork 34
Ohm Octave ~ A Musical Interval 34
Osteo Ohm Tuning Fork 34
Low Ohm Octave Tuning Forks 35
Practitioner Activator 35
High Ohm Octave Tuning Forks 35
Sonic Ohm Tuning Fork 36
CDs ~ Music in the Key of Ohm 37
Tuning Forks + Music 37

Part VI
Therapeutic Application
Acupuncture, Trigger & Reflex Points 41
Guiding Principles 43
Practice Application 43
Human Touch 44
Key Points 44
Holding Sacred Space 45

Part VII
Treating Self & Others

Self Treatment *(See List of Treatments on following page)* 48
Treating Others 50

Mid, Low & Osteo Ohm Tuning Forks
Application Process: Find, Palpate, Eye, Activate, Place 51
Application Methods: Single, Bilateral, Distal, Double 51

Body Mechanics 51

High Frequency Ohm Tuning Forks
Application Process: Activate, Direct 53
Application Methods: Rolling, Caduceus Weaving 53

Part VIII
Treatments
(See List of Treatments on following page)
Ohm Singing Bowls 76
Ohm Crystal Bowls 76
Contraindications 76

FAQs 77
Sources 79
Index 81
About the Author 87

List of Treatments

SELF TREATMENT

#1 Palm of Hand 48
#2 Top of Shoulder *(Shoulder Well – GB 21)* 48
#3 Bottom of Foot *(Bubbling Spring – Ki 1)* 49
#4 Sternum *(Primordial Child – Ren 17)* 49
#5 Sternum + Low Abdomen *(Ren 17 and Ren 6)* 50

TREATING OTHERS

#1 Listen 56
#2 Spinal Treatment *(Hua Tuo Jia Ji points)* 57
#3 Back of Head *(Heavenly Pillar – UB 10)* 58
#4 Lower Back *(Sacrum)* 59
#5 Back of Head + Lower Back 60
#6 Between Shoulder Blades *(Rhomboid Muscles)* 61
#7 Shoulder Blades *(Heavenly Gathering – SI 11)* 62
#8 Back of Legs *(Hamstrings and Calf Muscles)* 63
#9 Bottom of Foot *(Bubbling Spring – Ki 1)* 64
#10 Upper Chest *(Central Treasury – Lu 1)* 65
#11 Low Abdomen *(Sea of Chi – Ren 6)* 66
#12 Sternum *(Primordial Child – Ren 17)* 67
#13 Sternum + Low Abdomen *(Ren 17 and Ren 6)* 68
#14 Top of Shoulder *(Shoulder Well – GB 21)* 69
#15 Jaw *(Mandible Wheel – St 6)* 70
#16 Eyebrow *(Gathering Bamboo – UB 2)* 71
#17 Eye Region *(Fresh Innocent Eyes – GB 1)* 72
#18 Top of Foot *(Stream Divide – St 41)* 73
#19 Top + Bottom of Foot *(St 41 and Ki 1)* 74
#20 Energetic Field around Body *(Microcosmic Orbit)* 75

Charts

Guiding Principles .... *xxiii*, 43
Key Points 44
MID, LOW AND OSTEO OHM TUNING FORKS:
   Application Process: Find, Palpate, Eye, Activate, Place 26, 51
   Application Methods: Single, Bilateral, Distal, Double 51
HIGH FREQUENCY OHM TUNING FORKS:
   Application Process: Activate, Direct 27, 53
   Application Methods: Rolling, Caduceus Weaving 53
Comparison of Ohm and Schumann Frequencies 78
Electro Meridian Imaging Graph 40

OHM THERAPEUTICS GUIDING PRINCIPLES

1. A belief in the body's natural healing intelligence.

2. A knowledge that we are made of vibration.

3. A belief that applied vibration with Ohm Tuning Forks helps to remove blockages in the body's energetic pathways.

4. An understanding that an individual's biorhythms and cycles are intimately connected to those of the Earth.

5. A belief that sympathetic resonance with the Earth helps to restore balance, establishes homeostasis, and promotes healing.

PART I

Sound Healing

"We are each made of music."

~ Joseph Rael, Picuris and Southern Ute Author and Sound Healer

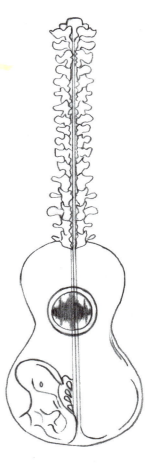

Sound Healing is based on the principle that everything in the Universe is made of vibration. Sound healing is possible because the human body is not solid and is held together by vibration. Like a well tuned musical instrument, the body is a rhythmic, harmonious form unless its vibratory field is disrupted.

Disease can be defined as any disorder, disruption or disharmony in the body. Sound, vibration and music have long been a part of our response to disease and illness in our quest for well-being. The inter-relationship between sound, music, health and the environment is inextricable. Novalis, the 16th century poet and philosopher, summarizes it best: "Every illness is a musical problem, and every cure has a musical solution."

For indigenous cultures throughout the world, the use of sound has been and continues to be an important part of ancient shamanic practices. Shamans employ sound in their healing ceremonies with their drums, flutes, bells, rattles, voices and other sound instruments, calling forth balance and creating harmony in earth cycles, and in the healing of individuals and their communities.

Sound Healing is all around us. Some people find that music, song and the spoken word have healing effects: the voice of a loved one, the sound of a child's laughter, a stranger's kind hello. Others experience healing in the vibrations that emanate from nature: a gentle breeze, the rhythmic waves of the ocean, a cat's familiar purr, a bird singing in the morning or crickets at dusk.

Our lives are busy and full and often our daily travels take us through noisy environments as we, for example, run errands, shop for groceries and drive in traffic. In our homes it is not unusual to hear the sounds of airplanes overhead, lawn mowers and leaf blowers, the neighbor's stereo system, and so forth. It can be difficult to turn off the unwanted noise, especially once those unwelcome sounds begin to resonate in our bodies. The stress produced by unwanted noise can cause our adrenals to overwork; it puts our bodies in a "fight or flight" state of alertness. This results in a state of imbalance and can create detrimental patterns of negative resonance or illness.

Listening to and experiencing the sound vibration of Ohm helps to dispel the unwelcome noise or resonance from our physical and energetic bodies. The grounding earth tone of Ohm helps the adrenals to relax enabling the parasympathetic and sympathetic nervous system to find balance. Once we clear the unwanted sound, we are able to find the quiet within.

Sound is the oldest form of healing; we have all experienced some kind of Sound Healing.

Sound vibration is at the heart of shamanic healing.

Sound waves enter the body where sympathetic vibration positively affects our cells, restoring healthy organization. Sound Healing is like a deep massage at the molecular level.

To understand the use of sound for healing, it is helpful to realize that all things vibrate: atoms, the cells of our bodies—all living matter—the Earth, her forests and oceans all pulse and vibrate with life.

Sound and music tend to bypass the mind and take us to a place of feeling.

Our bodies are largely composed of water, and water conducts sound.

Sound Healing works on the understanding that the body is not solid. Bodies are energy forms held together by sound or vibration. Disease can indicate we have gone out of tune or that the vibrational rate of the body has lost its rhythm.

Musical terms used in medicine:
- Get a "tune-up."
- to be in "disharmony."
- to be of "sound mind and body."

Benefits
- Promotes the flow of energy or "Qi" in & around body
- Opens the energetic pathways, alleviating stasis and relieving pain
- Relaxes muscular tension
- Relaxes adrenals, relieves stress and equilibrates the whole body
- Promotes sound sleep
- Promotes deep and balanced breathing
- Calming—facilitates balance and homeostasis in the body
- Stimulates the body's healing process
- Reduces joint pain and swelling by increasing natural anti-inflammatory compounds
- Promotes the healing of strained muscles, tendons and ligaments
- Increases bone density
- Energizing & revitalizing
- Enhances bodywork and energetic therapies, including Massage Therapy, Acupuncture, Acupressure and Shiatsu, Chiropractic, Polarity Therapy, Reiki, Therapeutic Touch, etc.

Our Vibratory System

The Ohm tone is an effective sound frequency to help neutralize the bombardment of external stimuli we experience in our daily lives. For example, the whirring of computer fans, the hum of electrical appliances, and the myriad of environmental sounds emanating from our neighborhoods. Our nervous systems begin to naturally resonate with all the unintended noise and sound byproducts in our home and work environments. Our adrenals respond by overworking and consequently we suffer from stress, lowered immune function, imbalance, and sleeplessness. The phenomenon of "vibing" or resonating with our surroundings happens to all of us. Another example of unwanted resonance is when someone else's mood affects or "rubs off" on you.

Our bodies are vibratory systems and we are naturally influenced by others and our surroundings. In a Sound Healing environment, we have the opportunity to clear and reset our sound equilibrium. The Ohm tone has a uniquely balancing and grounding quality, which makes it an ideal sound frequency to work with in the healing arts.

Our Earthly Body

Our physical bodies are made up of water and minerals. The closer we examine the organic nature of our bodies, the more evident it becomes that our physical body is a microcosm of the earthly body we inhabit. Our bodies, like the Earth, are primarily composed of water (70-80%) and we are also composed of very similar minerals and elements. Sound vibration travels extremely well through water. Because the human body is largely composed of water, it is a great receptor and conductor of sound. When we resonate with Ohm, we resonate with the Earth. To recognize the close affinity that exists between the human body and the earthly body upon which we live, is to understand why we would choose to resonate with Ohm.

The body is a perfect resonator for sound. Up to 75% of our body is made of water. Muscles are 75% water; the lungs are 90% water; blood is approximately 80 % water; and bones are approximately 25% water. Sound waves travel around 4 times faster in water. This makes the body an excellent receptor and conductor of sound and vibration.

Like the Earth and her many rivers, streams, lakes, and oceans, our bodies are traversed by networks of watery meridians. We are so much like the Earth in our physical composition that it makes perfect sense we would find ourselves reflected in her landscape. Photographic methods, electronic scanning and dark-field microscope techniques now enable us to see the most intricate details. Time and again, we see the microcosm is reflective of the macrocosm, from a molecular, cellular, and structural perspective: we are inextricably connected to the Earth.

Swedish Photographer Lennart Nilsson has taken a photograph of stem cells in bone marrow with the aid of a scanning electron microscope. The stem cells resemble clusters of wild pink roses. From another perspective one can see lichen or a coral reef. All points of view reflect the body's inner landscape. See photo in color section.

"As Above, So Below"
~ ANCIENT HERMETIC AXIOM

Intuition and Intention

There are parallels between a tuning fork placed on the body and a forked wooden branch used for intuiting subterranean sources of water. Both have a mysterious quality as they divine energetic pathways. The dowser and healer alike use vibration and resonance to run the channels or meridians in search of the energetic flow that courses through our individual and earthly bodies.

Intuition and intention go hand in hand. One balances the other: intuition is the yin and intention is the yang of a healing treatment. Intention brings the form of intuition to fruition. In life and in the healing arts, we gather information and guidance when we tap into the spring of inner knowing. Knowledge, the intuitive process and clear intention all contribute to a healing treatment.

Both dowser and practitioner use vibration and resonance in their practice. The dowser locates channels of water flowing through the Earth, and the healer finds the energetic channels present throughout the body. In the practice of water divination, it is the forked branch which resonates; perhaps the xylem and phloem within the wood resonate with the watery meridians beneath the surface of the Earth. This ancient practice was intuitive and directed by the dowser's familiarity with a landscape and his or her intention. The use of tuning forks on the body is governed by the same idea: an understanding of resonance, knowledge of the body, intuition, and intention.

Resonance and Entrainment

Resonance is a sympathetic vibration between two objects or bodies. For example, when one Mid Ohm Tuning Fork is activated and held in proximity to another Mid Ohm Tuning Fork, the unactivated tuning fork will begin to vibrate. This demonstrates the principle by which one like body or vibration is drawn to another, causing both to share a sympathetic resonance.

Entrainment is similar to resonance. Physicist Christian Huygens coined the term entrainment after he noticed, in the 1600's, that two pendulum clocks had moved into the same swinging rhythm with one another. This example of entrainment shows the process whereby two vibrating systems fall into synchrony.

Sympathetic Resonance

Sympathetic Resonance occurs when one vibrating object influences another object to also vibrate. Previously, an example was given of how an individual's mood can be affected by the disposition or behavior of another person. Sympathetic Resonance describes what happens when our feelings and behavior are influenced by another person or our environment. Activities such as listening to music or being in nature can change how we feel, demonstrating the field of influence of that activity. For some, eating food that is mindfully grown without pesticides, from organic seed in soil that is alive with beneficial organisms, is a conscious choice to sympathetically resonate with a positive life affirming healing vibration. By becoming aware of the forces that influence us, we are better able to understand the concept of sympathetic resonance.

Women's menstrual cycles can correspond to lunar cycles, and women within a group often find they begin to cycle at the same time each month. This example demonstrates resonance with others and our environment, in particular, through the influence of the moon. It also demonstrates the concept of entrainment, as shown by the synchronization of the menstrual cycles of members within the same group.

When individuals are put into isolation, dark rooms that disconnect them from the influences of a natural environment, they lose a sense of time and place. This disconnect from natural cycles, such as sunrise and sunset, creates a loss of center and balance, resulting in disorientation. We are deeply connected to earth rhythms and cycles, and loss of this vital connection can lead to physiological and psycho-spiritual illnesses. In a state of good health, we resonate with the Earth and are entrained to her rhythms and cycles. Examples of this positive resonance or entrainment include regular and uninterrupted sleep cycles, a balanced pulse and heartbeat, regular bowel movements, healthy respiration and appetite.

The vibrational sound waves generated by Ohm Tuning Forks establish a healing resonance in our bodies, helping to replace any negative resonance or disharmony that may be contributing to illness. This healing resonance or "body attunement" is calming, soothing and relaxing and can be revitalizing and energizing.

When we experience the sound vibration of an Ohm Tuning Fork and/or music composed in the healing key of Ohm, we experience sympathetic resonance with the seasons, cycles and the Earth's life force.

Because we share the same relative percentage of water and minerals, our physical makeup is very similar to that of the Earth. *Ohm Therapeutics* is a healing system based on the premise that resonance with the Earth helps restore balance, and re-establishes homeostasis, vital to the healing process. It is widely accepted that our individual biorhythms and cycles are intimately connected with those of the Earth. As we heal ourselves through resonance with Ohm, we also have an opportunity for sympathetic healing with the Earth.

WINTER SONG

Painter Donna Caulton captures the musicality of Earth through her visual depiction of cycles, seasons and moon phases, themes all present in her series of resonance paintings.

Our Earthly Body

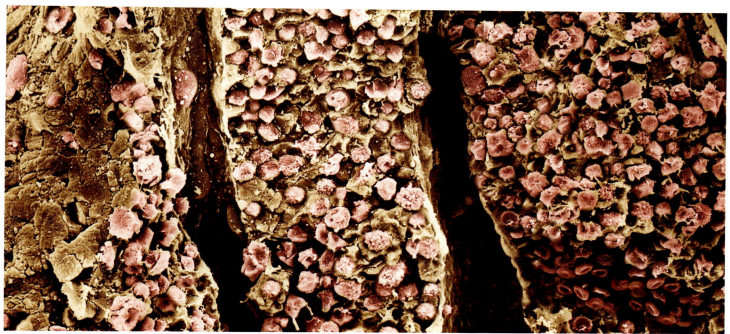

With the aid of a scanning electron microscope, Swedish Photographer Lennart Nilsson voyages within the human body. "This is bone marrow," says Nilsson. "Some of the red round spots are stem cells that are going to develop." This astonishing photo of the inner body landscape resembles lichen found on rock, a corral reef, even a pink cluster of wild roses. PHOTO ©LENNART NILLSON

A double rainbow over Glacier National Park demonstrates the concepts of overtones and octaves. PHOTO ©FOREST WOODWARD.

A Solar Year is symbolized through a rendering of the Earth orbiting the Sun. PHOTO ©MANUELA MILLER

RESONANCE

In her painting Cadence, *artist Donna Caulton depicts the musical elements and rhythms present in nature.*

From the popular television series Bewitched, *Doctor Bombay treats Samantha for a malady using a tuning fork.* COURTESY SONY PICTURES TELEVISION.

Tuning Fork Applications

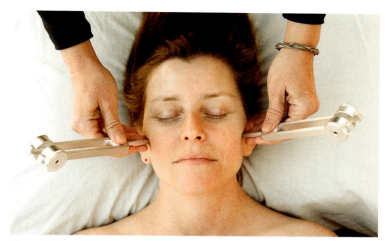

Sound Facial™ Treatment

Relax, ground and center with Ohm Tuning Forks during a yoga class at Body, a Health Center & Spa in Santa Fe, New Mexico.

Sound Healing with Ohm Tuning Forks therapeutically enhances On Site Chair Massage.

A Mid Ohm Tuning Fork applied to certain Acu-points will help you go the extra mile while hiking and enjoying other outdoor activities.

PHOTOS THIS PAGE ©WENDY McEAHERN

Sound Healing Tools

Ohm Crystal Bowl

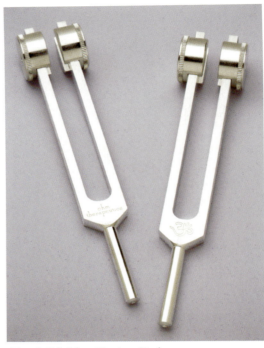

Two Mid Ohm Tuning Forks

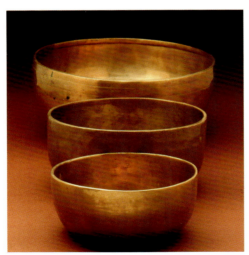

Ohm Singing Bowls

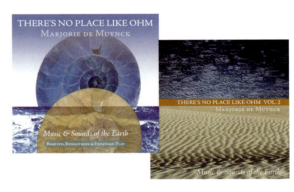

Music in the Key of Ohm

PHOTOS © WENDY McEAHERN

PART II

Sound & Vibration

"Sound is the hearing part of Vibration and Vibration is the feeling part of Sound."

The emerging field of Harmonic Medicine includes Music Therapy and Sound Healing. Sound Healing practices incorporate a variety of vibrational tools and instruments, often featuring the therapeutic application of tuning forks, singing and crystal bowls, gongs, bells, sound plates, rattles, drums, didgeridoo and the voice. When listening to these instruments and various forms of vocal expression—singing, chanting, toning, Tuvan throat singing, overtone singing from Tibetan monks—one can hear and feel the crossover between sound and vibration.

Vibration and Form

Everything that exists is a testimony to the vibration that gave it form, from the five pointed shape of a starfish and the petals characteristic of many flowers, to the peaks and valleys that shape our own bodies. Without a doubt, the drone of bees and the birds' song are frequencies that create their own unique and sacred sounds, shapes and geometries.

Think of the body's pulse as a wave generator. These waves create vibration and resonance to collectively form who we are. Rainbows, whirlpools, water droplets, sand dunes, ocean waves, and the wind blowing through fields of wheat, all are examples of vibration creating form.

As a way of demonstrating this phenomenon, Hans Jenny developed *Cymatics*. This study of wave theory shows that vibration creates recognizable shapes and forms. In a controlled environment, various frequencies of vibration were applied to a tonoscope. A tonoscope is an electronically controlled plate upon which a medium (a fine grained substance) is placed. As the plate vibrates from applied frequency, the medium takes on various configurations. What Hans Jenny demonstrated in the science lab are the manifest shapes and forms of the natural world.

> Hopi and Navajo traditions tell of a time when Shamans would speak or sing into sand paintings during ceremonies and healing rituals, creating sacred patterns. In the Hindu tradition these vibrational representations are called mandalas.

OHM YANTRA. Yantra is a Sanskrit word meaning "to sustain." A yantra is the geometrical representation of a mantra or sacred sound, in this case Ohm.

The natural world is a tonoscope palette. Anyone who has a garden observes the vibrational hum of summer as the abundant energy of growth takes shape as fruits, vegetables, herbs and other offspring. Another example of a vibratory force in nature is the repetitive ebb and flow of the ocean as the surf continuously washes over changing the shape and form of the land.

Sound is energy; it is the basis of form and shape.

The laws governing Quantum Physics demonstrate that everything in the Universe is made up of vibrational fields, including the human body.

Sound vibration communicates with the body. Every cell is made of vibration.

Humans, animals and plants—all life forms communicate with one another via sound.

Vibrating bodies respond to sound. When an object, such as a tuning fork, is activated it begins to vibrate. The vibration pushes the surrounding particles, causing them to move, creating waves of energy, which emanate from the tuning fork. This sound wave is called kinetic energy.

Sound waves are three dimensional.

Sound waves that are closer together produce a higher frequency or pitch.

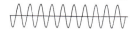

Sound waves that are farther apart produce a lower frequency or pitch.

Both high and low frequency vibration possess incredible therapeutic potential.

Frequency is determined by how fast or slow the waves move, and is known as cycles per second (cps).

The Mid Ohm Tuning Fork has a healing frequency of 136.10 cps. The term "hertz" (abbreviated hz) is interchangeable with cps and is more commonly used to designate tuning fork frequency.

Sound and Vibration create recognizable shapes and forms.

The vibratory essence of sound affects the inner walls of nerves and blood vessels.

Sound and Vibration can strengthen the cells and tissue, and discourage unhealthy cells.

Electromagnetic Frequencies

Our sensitivity to the subtle healing influences of the natural world is in decline. We are increasingly desensitized, in part, due to living in a world excited by frequency from computers, wireless communication devices, power transformers, satellite transmissions, etc. At home and work, we are continuously bombarded by noise and electromagnetic frequencies (EMFs) that disrupt our natural biorhythms and cycles. For many of us, noise pollution and over stimulation are common complaints and contribute to a constant low-level anxiety, affecting our ability to deeply relax and sleep well. Without relief, any form of stimulus can overtax our nervous systems.

AC or 'alternating current' fields are electrical emission byproducts of human enterprise and are very different from the Earth's magnetic field. For example, the power that illuminates the lights in our treatment rooms, heat lamps and heating pads, stereo systems, and so on, all emit an energy field. Many of these devices are necessary and greatly improve our quality of life. However, there is a tipping point and we can have too much of a good thing. Some studies that suggest AC magnetic fields can have profoundly negative effects on human cell behavior. Damage from EMFs is a real form of pollution, and is increasingly quantifiable. An awareness of these influences in our environments is imperative.

We are just beginning to learn of the unintended health consequences from our exposure to these EMFs, known as Electro Magnetic Radiation (EMR). For this reason, introducing electronic and frequency-emitting devices into a sound healing environment is questionable. Further, we risk losing our sensitivity to the subtleties of treatment when we begin to rely on machines.

THRESHOLD OF NOISE

While composing the CD In the Key of Earth, *I became acutely aware of the threshold of noise that has become, for the most part, our fundamental or base level frequency.*

Cars, buses, power transformers, planes, televisions, etc., all contribute to the daily envelope that makes up our sound environment. The threshold varies, but the din of noise pollution that surrounds us is clearly on the rise.

I wonder why Western music consistently tends to be so full and loud, with extremely high volume refrains toward the ends of musical pieces. Even some soft rock and new age music have raised volume thresholds to accommodate this increase in our base level frequency.

My observation is that we continue to raise the decibels (volume) of our music in order to offset the unwanted sounds in our environment. Perhaps we are trying to be heard above all the machine noise. It is become increasingly more difficult to hear the rich rhythms that occur in nature, along with the sounds of our own rhythmic breathing and heartbeat.

Medical Application of Frequency

Sound and Vibrational Healing can effectively restore balance to the body. The root of illness can be seen as the disruption of harmony in the body, which results in negative resonance or disease.

The restoration of balance and harmony enables the body's immune system to function more efficiently. Enhancing and strengthening the overall body, including the immune system, is the primary objective in the Ohm Therapeutics model of Vibrational Sound Healing. The premise of this work is to support the body's natural healing ability. The use of acupuncture points to access Qi and the application of sound frequency known for its ability to move Qi help to regulate and strengthen the whole body, achieving homeostasis.

The use of Vibration and Sound Healing are increasingly of interest in Western Medicine. Allopathic medical practices employ vibrational technologies such as ultrasound and magnetic resonance diagnostic tools. Dental offices use sonic tools to clean teeth, and lasers are being used for some surgical procedures now as they offer less bruising, bleeding and pain. Music is being introduced into hospitals to further relax patients while undergoing surgery —as well as for the benefit of the doctor performing the surgery—and Music Therapy is widely known to assist in the patients' recovery.

In the diverse field of Complementary Alternative Medicine, Vibrational Sound Healing is being incorporated into a variety of healing modalities. Questions often arise concerning which frequency to apply to which health condition. When applied in this manner, frequency is at risk of being labeled and prescribed in much the same way as pharmaceuticals and herbal remedies. The caution here is in treating the illness, and not the individual.

Treatment protocols are being offered that direct specific frequencies to specific body organs with the intention of healing that organ. Some assert that a certain tone or frequency is associated with a particular organ or, in respect to the subtle body, a specific chakra or energy vortex. Based on this premise the practitioner applies a specific tone or frequency to that organ. This method of treatment strongly resembles Western practices.

While there may be a similar range of frequency for the liver, all livers do not vibrate at the same rate, nor do they have the same frequency from minute to minute, or day to day. Those familiar with the circadian clock and biorhythms of the body understand this to be true. By treating holistically, the practitioner facilitates the body's inborn ability to self regulate and heal that which is impaired.

Applying a specific frequency associated with a specific organ is not unlike prescribing a drug. Treating from this viewpoint is based on the assumption that a particular condition manifests the same for everyone. This is at its best a "hit and miss" medicine. For many years chemotherapy was broadly prescribed for anyone with cancer, with very unfortunate results for those unable to tolerate this type of treatment. An individual can now be tested to determine if they are a good candidate or not for specific types of cancer treatments.

> High frequency is known to shatter glass. A television commercial for Memorex in the late 60's shows Ella Fitzgerald hitting a high note and breaking a crystal wine glass. In the television series, *Star Trek,* there is an episode in which intruders use high frequency to inflict pain aboard the Enterprise; Captain Kirk and his crew drop to their knees and cover their ears in agony.

> The US Navy uses intense frequency sonar in military exercises which are known to injure and kill whales and other marine life. Active sonar systems produce waves of sound that sweep the ocean like a floodlight, revealing objects in their path. The decibel levels of some systems can spread harmful sound across hundreds of miles of ocean.

Frequency is powerful and its effect is palpable. There are studies that indicate frequency applied to cancer cells can cause them to "blow up" or become destroyed. At first glance this information appears hopeful and exciting. However, this seems to be a hap hazardous and dangerous method which may further spread the cancer. A tumor is an example of the body's intelligence as it walls off and contains malignant cells. A warfare approach is still very one dimensional: it focuses on treating the manifest symptom(s) of the underlying illness (usually through aggressive measures and rhetoric) rather than strengthening the whole person and medically addressing the possible causes for the cancer or illness itself.

As Sound Healing re-emerges into contemporary medicine there are bound to be many diverse and sometimes conflicting viewpoints. Understanding the guiding principles and rationale for the selection of frequency is of key importance in any Sound Healing system.

Ohm has proven to be a very effective sound vibration to facilitate homeostasis and helps to restore balance and harmony to the body. This homeostasis allows the body's innate healing inteligence to do its job. It is thought that during an earlier time in our history, the sound of the Earth's rotation and revolution around the Sun could be heard; this phenomenon is known as "Music of the Spheres" in the Occidental world. The elliptical path of the Earth around the Sun is the music of our sphere, its vibrational influence permeates all of life. Experiencing Ohm connects us to something greater than ourselves while energetically rooting us to the Earth and attuning us to the beneficial forces of Nature.

PART III

Ohm, Octaves and Overtones

OM

"The imperishable sound,
Is the seed of all that exists.
The past, present, and the future,
Are all but the unfolding of OM.
And whatever transcends the three realms of time,
That indeed is the flowering of OM.
This pure self and OM are as one,
And the different quarters of the Self
Correspond to OM and all its Sounds."

~ Mandukya Upanishad

In Hindu cosmology, Ohm is considered a source tone. Known as *bija* Ohm is a seed syllable used in mantras. According to ancient Sanskrit religious texts, the syllable, mantra and vibration of Ohm is said to be the primordial sound of the Universe. For this reason, the Ohm vibration is thought to contain all sound, just as white light contains all color. Ohm is considered sacred and healing; its sound vibration and iconic symbol have been represented in spiritual practices and interpreted musically for thousands of years by cultures throughout the world. Other spellings of Ohm include Aum and Om, and the phonetic spelling Aoum.

ohm

In this system, Ohm is spelled with an "h." Traditionally the "O" represents birth or beginning and the "M" represents death or completion. The added "h" in our spelling for this symbol provides a visual axis mundi, connecting spirit and matter. The phonetic addition of the "h" reflects our desire to emphasize the element of air, and thus breath, vital to life and the healing process.

Ohm is an ancient and beautiful symbol with many stylistic variations. The Sanskrit symbol is likely the most widely recognized, followed by the Tibetan symbol.

Historically, Ohm symbolizes the totality of life, from birth to death, the waking and dream states, and our corporeal and transcendent selves. Through time, the ultimate mystery and message of Ohm continue to be explored. Respectfully, *Ohm Therapeutics* contributes to the evolution of this beautiful symbol by including an image of the Earth, our home.

Ohm and C

The musical note C (128 hz) is central to Western Music. It is in the center of the piano keyboard; the C scale is played on all white keys, with no sharps or flats (black keys). Mid Ohm is very close to C and C# at 136.1 hz, and while close to one another, musically they are worlds apart.

There is conjecture that several hundred years ago the church of the Western world and its representatives decided to ban certain tones and/or overtones. Dissonance was not acceptable in music for worship. A tone associated with the Earth, such as Ohm, and its naturally occurring overtones, might be considered Pagan because of its association with nature, astronomy and Omnipresence.

It isn't a far stretch to consider that the church established musical center of C was purposefully shifted away from Ohm. Ohm is slightly more than a semi-tone from C. When the system was set up for the tones or musical notes to be equally distanced from one another (e.g., on the piano) C became the new center. The system of placing intervals at equal distances to one another is called Equal Tempered. Our current Western music has its origins in this European system which is centered around the piano and the musical note of C.

Music theory, with its system of musical intervals, was most likely born of what was naturally heard in overtones. Just as we have selected and named primary colors from the rainbow, particular tones have been chosen to create a musical system from overtones. The octave is prominent in the overtone series, followed by a beautiful spectrum of subsequent healing tones, each unique and restoring. Great musical and healing potential exists in overtones, in their manifold shades and variations of color and sound.

Ohm ~ Our Musical Center of Gravity

In music a fundamental key defines the musical piece. The same holds true for Sound Healing, where the key roots or sets the tone. In this vibrational healing system Ohm is that tone. When asked by students, Why Ohm? We answer, *because* Ohm is an earth tone. Its calculated frequency is scientifically based and its sound vibration is spiritually upheld throughout the world. Resonating with the Earth is beneficial—it is our home tone.

In the Ohm Therapeutics system, the measured frequency of 136.1 hz is the fundamental tone and is called "Ohm" or "Mid Ohm." Other words used to describe Ohm as our tonal center include the following: root, tonic, key and keynote. Ohm is based on the measured frequency of the Earth's elliptical orbit as it travels around the Sun through four seasons. The Ohm tone is reflective of the Earth's relationship to the light of the Sun and is sometimes called a Solar Year. Each season brings forth the elements. This yearly cycle, characterized by the rhythmic flow of the seasons, lunar cycles, and the pulse of day and night, is what gives the Ohm tone its richness in the medicine wheel of life.

When an organism is out of balance, it becomes weakened and vulnerable. Resonating with Ohm positively affects our biological rhythms and circadian clock. We begin to sync or "entrain" with natural cycles which enable us to find our balance. Balance and homeostasis are essential to our health and create a positive environment for healing to occur.

Ohm is clinically shown to be a safe and therapeutic sound frequency. Because the Earth is our home and gives us life, the fundamental tone of Ohm is an excellent choice for healing work.

Mid Ohm Frequency ~ 136.10 hz

136.10 hz is a numerical representation of the vibrational tone of Ohm.

Astronomer Johannes Kepler, influenced by his musical upbringing, realized in the seventeenth century that planets have elliptical orbits. This gave basis for the modern calculation of planetary orbits and their associated musical tones or frequencies.

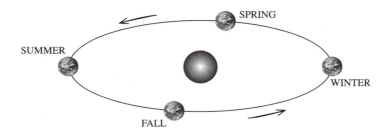

The musical tone of Ohm, once intuited by ancient cultures, owes it modern calculation to Swiss scientist and astronomer Hans Cousto. The calculated frequency of Ohm is raised appoximately 32 octaves, to arrive at 136.1 hz, to be audible for human hearing. Om, Ohm or Mid Ohm all refer to the measurement of 136.1 hz. In Ohm Therapeutics, we use the term "Mid Ohm" because it is at the center of our vibrational healing system.

Human conversation ranges between approximately 200-400 hz. Significantly, the Mid Ohm is just under this range at 136.1 hz. This lends to its tonal familiarity, and ability to provide a sense of relief and calm, without deadening our senses or creating a feeling of being "spaced out" which can be experienced in alpha state sound and music.

Keynote of Ohm

Ohm is the fundamental tone found in the music of many ancient cultures. For example, in India, the fundamental tone of the sitar was traditionally tuned to the *sadja* or musical tone of Ohm. The sitar was tuned to the tone of the earth year, a harmonic based on the Earth's cycle around the Sun.

It is customary for a musician to tune the sitar prior to performing; this prelude or pre-tuning known as the *alapa* is said to be very beautiful. During the alapa, the instrument (sitar), musician and audience become attuned to Ohm. Musical instruments tuned to Ohm are found elsewhere in the world. Other examples include ancient Irish and Indonesian musical instruments as well as ancient Tibetan singing bowls and bells.

CONTEMPORARY OHM TUNING—SINGING BOWLS

When Lemniscate Music went in search of Singing Bowls tuned to Ohm we discovered how rare it was to locate newly made bowls with Ohm as the fundamental tone. Working closely with a colleague, who listened and measured the frequency of each bowl in her shipments from the East, we found that an extremely small percentage of these singing bowls were tuned to Ohm. Together we began to work directly with Himalayan artisans to revive the old practice of actually hand-hammering and tuning the bowls to this traditional and sacred frequency.

At the same time, another colleague in Indonesia working with musical artisans asked them if they recognized Ohm, demonstrating the tone with an Ohm Tuning Fork. An elder recognized the sound and shared that Ohm was known in a previous time. With the modernization of the world, this tone is slowly becoming lost to us. The search for Ohm and recording of its influence and significance in other cultures continues.

Resonance with Ohm
Intuition, meditation, and a connection to Nature through an observance of the Earth's rhythms and cycles led individuals in the past to resonate with the Ohm tone. Examples of this include the daily sunrise and sunset, eclipses, and seasons of the year. Ohm is found in ancient religious texts, the seed syllable of mantras, and in the tuning of traditional instruments. Furthermore, science has confirmed that Ohm is the measured frequency of the Solar year.

> *"The earth tone of Ohm helps connect us with something natural, sacred, and familiar."*

Many express a feeling of deep recognition when listening to Ohm. At the beginning of a new class, students are asked to describe what they experience when listening to the Ohm tone. Their intuitive responses show how we all are connected to one another, the Earth and our place in the wider cosmos. The vibrational tone of Ohm elicits positive and life-affirming associations. The following words and phrases were offered by students after they experienced the sound vibration of an Ohm Tuning Fork, often for the first time.

> cosmic hum, mother love, heart of the universe, communicating with the universe, pain relieving, nurturing, calming, ancient knowledge, primordial sound, bumblebees, peace, comfort, safe, embraced, energy dance, coming home, familiar, flowing, tension relieving, breath deepening, a happy heart, a sense of the "all knowing," moving in the infinity, a garden…

Octaves

The term octave comes from the Latin *octavus*, or the female derivative *octava* which means eight (8). On its side, the 8 represents infinity, and upright, the number eight suggests the axiom "As Above, So Below."

The Lemniscate Music logo represents the octave and is a playful depiction of the Ohm Muse. The logo embodies the shape of an octave in the form of the musical staff beautifully draping the Muse's body.

The octave is omnipresent in the natural world. It is an earthly and cosmic phenomenon found in mathematics and science as well as in music and poetry. Examples of octave duplication or repetition are found in the unfurling patterns of plants and mineral formations—the nautilus for example—and in the shapes of galaxies. According to Theosophist Helena Blavatsky, eight, the number of the octave, represents the eternal, spiral motion of cycles.

The double helix of DNA and the Caduceus are made up spiraling of patterns of eight built on top of one another.

Octaves are determined mathematically by multiplying the root frequency (e.g., 136.1 hz) times two, for the next octave up (277.2 hz), or by dividing the root frequency in half, for the next octave down (68.05 hz).

In music, an octave refers to a series or range of eight notes or tones including the root and the octave which have the same name. In poetry, an octave is a rhythmic group of eight lines of verse. Octaves are also present in cultural practices. The seven days of the week are like the seven tones of a musical scale, where the eighth day or tone begins a new cycle: the eighth is the octave. Sunday is same day of the week but a further place in time. Do is the same musical note only at a higher or lower octave. An octave is the distance between two like tones.

Do	Re	Mi	Fa	So	La	Ti	Do
Sunday	Monday	Tuesday	Wednesday	Thursday	Friday	Saturday	Sunday
1	2	3	4	5	6	7	8

The octave is the primary overtone. It is the closest tone to the fundamental and the simplest numerical ratio to the fundamental. An octave is the most prevalent and vibrant interval in music and in nature. All tone occurs within the octave.

Octaves are found in shades of color. Lighter shades of blue are higher octaves than their darker counterparts. Octaves occur in the rings of trees, in weather patterns, and in molecular vibratory phenomena.

In 1863, English chemist John Newlands discovered the Law of Octaves by observing that natural elements with similar physical properties always differed in atomic weight by eight (8). This was integral to the formation of The Periodic Chart of the Elements.

In Sound Healing, the octave is tremendously effective because it creates movement along the axis of a root tone, in this case Ohm, providing latitude to move up and down, with higher and lower octaves. The healing aspect of the octave exists in its affirming repetition, and the movement it creates. It is noteworthy that the octave is the most frequently occurring interval in the overtone series.

> *"The body in its natural intelligence is able to assimilate Ohm and all its beautiful overtones, and utilize these healing properties on a vibratory level."*

Overtones

In the realm of sound and music, an overtone is a harmonic that naturally occurs from a fundamental sound. Playing a didgeridoo, singing, toning and chanting are all activities that produce overtones. Overtones also result from motors and engines, a cat's purring, the thrum of insects, plucked guitar strings, hammered metal—sounds we hear everyday.

In music and in sound, for every frequency or sound there exist subsequent tones. These sounds are called overtones. Overtones are a naturally occurring resonance and follow in an intervallic series from the fundamental tone.

A rainbow is one of the best visual examples to demonstrate the phenomena of overtones. A rainbow is an arc or bow of prismatic colors that appears in the sky. Within one primary color band you will see overtones (various lighter shades) of that color. Consider the primary band of color as the fundamental. See photo in color section.

Overtones are always present, even though they may not be audible to the human ear. The most frequently occurring overtone is the octave. When you activate and listen to the Ohm Tuning Forks, you may hear and/or experience their resulting overtones through listening, and through skin and bone conduction.

The therapeutic potential of the overtones of Ohm is tremendous. The sound vibration of the root tone of Ohm and its overtones reach the body through the resonant chambers of the ear and through bone, which are both excellent resonators because of their hollow structure and high water content. Each person's body will create different overtones based on their physical and energetic make-up, overall health, and individual needs at the time. The body is like a landscape. Imagine sound is like water that is naturally drawn to and absorbed first by the dry or "deficient" areas within that land, bodyscape or bioterrain.

With respect to tuning forks, overtones are the resulting unstruck tones that occur when a tone is struck or activated. For example, the Low Ohm Tuning Fork (68.05 hz) produces a low struck tone, and subsequent higher tones. These higher tones are overtones. In some cases, the overtones are so abundant that it makes it difficult to hear the fundamental tone! This is often true with quartz crystal and Tibetan-style singing bowls.

PART
IV

Tuning Fork Technique

"The beauty of the Ohm Tuning Fork is its versatility—it is non-invasive and lightweight, and each time you activate the tuning fork it rings anew with the vibration of Ohm. It is a totally 'live' experience."

Tuning Forks are lightweight sound healing tools. When activated their vibration is easily directed on the body, and in the energetic field around the body. With tuning forks, sound energy is uniquely and effectively transferred to the body, making them easy tools to incorporate into a wide variety of healing modalities. Another advantage is portability: tuning forks are easy to handle and travel well.

Disharmony can manifest in the body as stress, tight and sore muscles, and fatigue, creating blockages to our Qi or natural energy flow. These blockages can lead to illness. The sound waves created by the Ohm Tuning Forks work like kinetic energy to move disharmony and tension from the body, remove Qi stagnation, and help to restore a sense of balance and well-being.

Ohm Tuning Forks are recommended for use on tight muscles, joints, tendons, bones and tissue, as well as on acupuncture (acupressure), trigger and reflex points. Tuning forks are non-invasive and fully able to engage the Qi. Sound vibration easily travels the body's meridians and neural pathways and enhances other bodywork therapies.

Acoustic Advantage
Ohm Tuning Forks are lo-tech acoustic musical instruments; they produce sound vibration through simple activation, without electronics. This is an added bonus. Working with tuning

forks allows us to employ a healing apparatus that has no additional frequency byproduct. The purity of the intended tone (and its subsequent overtones) creates a sound environment that facilitates homeostasis and is conducive to healing.

Parts of a Tuning Fork

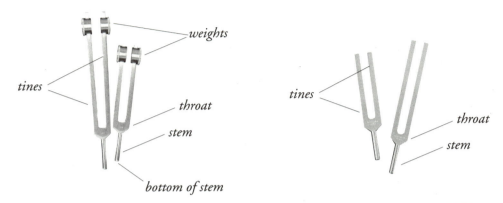

A question that often arises concerns the tuning fork weights, and why they are present on the mid and lower frequency tuning forks. The weights add mass, which lowers frequency. Thus, a greater amount of energy is transferred to the stem of weighted tuning forks, which is ideal for applying and directing sound vibration onto the body.

The length of a tuning fork is another factor in determining its frequency. Other factors include width, thickness and elasticity. Since the width, thickness and metal are generally constants, the frequency is changed by making tuning forks different lengths. As a general rule, the greater the mass of a tuning fork, the lower the frequency. Longer tuning forks have greater masses and lower frequencies. In this system, the Low Ohm and the Osteo Ohm Tuning Forks are examples of this dynamic.

Learning how to correctly hold and activate a tuning fork will greatly enhance your practical application. It is not uncommon for those first handling tuning forks to grip the forks too tightly, causing their hand to fatigue and cramp. How you hold the tuning fork will affect your ability to comfortably place it on the body. Once the correct techniques are learned, it will free you to creatively explore tuning fork application.

How to Hold a Tuning Fork

Hold the stem of the tuning fork with a firm yet relaxed grasp between your thumb and fingers. Spread your fingers slightly apart to avoid creating tension in your hand. If the fork is held too high up on the tines it will dampen the vibration, and if held too low, it could cause cramping in your hand. Your forefinger and thumb should hold the tuning fork at the throat where the stem and the tines meet.

How to Activate a Tuning Fork

Take a deep breath, relax your hand, wrist and arm, and strike the grooved edge of the weighted end of the Ohm Tuning Fork on the Practitioner Activator. Use the motion of your wrist to activate the tuning fork, as opposed to using your entire arm and shoulder. Make sure you activate the tuning fork with enough energy to provide a strong and lasting vibration.

Wearing a Practitioner Activator

The Practitioner Activator is designed for those who work in therapeutic settings. An adjustable strap fits around your thigh, and allows you to freely move around a table while re-activating the tuning forks during a treatment. Most therapeutic applications require two tuning forks, for working distally and bilaterally on the body, and for doubling the healing resonance when needed. Wearing the Practitioner Activator leaves both hands free for tuning fork application. Some therapists choose to wear two Practitioner Activators, one on each leg, finding it better facilitates good body mechanics and adds to the pace and rhythm of treatment.

To maximize the sustain (length of vibration) and minimize the sound of the "hit" on the activator, strike the tuning fork at an angle on the ribbed edge of the weight. Keep your shoulder, arm and wrist relaxed as you activate the tuning fork. Practice activating the tuning fork so the motion is rhythmic and feels comfortable.

How Long will a Tuning Fork Vibrate?

The vibration of a well-activated tuning fork can sustain for up to 20-30 seconds. Factors influencing the length of vibration are the location of application and the individual's needs at the time of treatment. If the vibration dampens quickly, check to be sure you are holding the tuning fork correctly. Another explanation for the tuning fork losing its vibration quickly could be that the body is rapidly taking in the sound. Earlier, the body was described as a landscape that gathers moisture where needed; sound is like water being drawn to the dry (deficient) areas. When an area is saturated with water, the wet ground will not continue to absorb as much or as quickly as if it were dry ground. The same concept applies to how sound is absorbed by the body.

The speed at which the vibration is absorbed can be thought of diagnostically. In some areas the sound vibration will be rapidly absorbed. These areas of the body will benefit from additional sound application. With repeated application, the muscle and tissue will begin to soften or relax. Now the sound vibration will be sustained for a longer period, as the body is now measurably less 'thirsty' or deficient in the area being worked on.

Number of Applications

The individual's needs at the time and the area of the body being treated will guide your decision regarding the number of applications. Some places will need only one application or pass, while others may need more. If you are not sure, then apply the tuning fork three times. After the vibration has ended, sound is still traveling through the body. Some people report they feel the sound for several hours to a few days after a treatment. Your body in its natural healing intelligence will utilize the sound it needs. This is similar in concept to the body's natural ability to intake from water soluble vitamins and minerals only the amount it needs, eliminating the rest.

Applying the activated tuning forks 2-3 times to the same area or point on the body is usually sufficient; however, more applications are appropriate if the body continues to rapidly absorb the sound in that area. With practice you will be able to gauge the effect of the sound vibration and begin to sense how often to apply the tuning forks to the body.

MID, LOW & OSTEO OHM TUNING FORKS

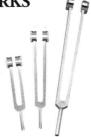

APPLICATION PROCESS: (on the body)
1. **FIND** the location for desired application.
2. **PALPATE** the location.
3. **EYE** the location.
4. **ACTIVATE** the tuning forks, while eyeing the location.
5. **PLACE** the tuning forks on the body.

Applying to the Body: Mid, Low & Osteo Ohm Tuning Forks

When treating yourself, breathe deeply with your body in a relaxed position. Start by listening to the activated tuning forks, holding them just a few inches from each ear, as feels comfortable. When you are treating someone else, check in with that person and ask if they can hear the vibration comfortably. Adjust the tuning forks to find the comfortable distance from their ears. It is recommended that you start each session with listening.

When a fetus is approximately four months in utero, it begins to hear sound and vibration. The fetus will turn its head to locate the sound it hears. Hearing the sound stimulates a sense of proportion and balance and is essential to development.

Listening to a familiar sound in a gentle and balanced manner is an effective way to begin a Sound Healing session.

Next, palpate the point or muscle where you want to place the tuning forks, eye the location, then in a smooth and unhurried motion "activate" and bring the tuning forks to the body. With practice, the activation process will become a natural and easy motion. Keeping your eye on the location will help direct your hands as they hold the vibrating tuning forks and bring them to the area being treated.

Be sure that both the tuning fork and the edge of your hand touch the body. This contact helps to relax your hand and comforts the recipient with the warmth of human touch while applying the tuning fork. See DVD.

Now apply the bottom of the stem of the activated tuning fork to the body with the intention of establishing a deep energetic connection. It's important that you use enough pressure to ensure that the vibration of the tuning fork is clearly felt by the recipient. Press hard enough or the vibration might feel like tickling on the skin.

Treat both sides of the body for balance. Use your good sense; and check in to make sure that you and the individual receiving the treatment are both comfortable.

HIGH FREQUENCY TUNING FORKS

APPLICATION PROCESS: (energetic field around body)

1. **ACTIVATE** the tuning forks in a rhythmic manner, first one and then the other, and gently enter the energetic field around the body with the intention of clearing negative energy, and strengthening this field.

2. **DIRECT** the tuning forks above and around the body, keeping them somewhat parallel to the body, as demonstrated on DVD.

Tuning forks can also be directed toward the corners of a room to move stuck energy, and to balance the energy of the entire space.

Applying High Frequency Tuning Forks to Energetic Body

The High Ohm Tuning Forks are ideal for use in the energetic field around the body, and for clearing and resetting the energy of a room. The sound wave produced by the High Frequency tuning forks is therapeutic and can help restore balance to the energetic field around your body, thereby benefiting your physical body.

The method for activating the High Ohm Tuning Forks is the same as for the Mid and Low Ohm Tuning Forks. Wearing a Practitioner Activator will enable you to strike the tuning forks on a rubber surface that has been chosen for its properties of thickness and hardness (durometer). The durometer of rubber used for the table top Activator and the Practitioner Activator provides a stable and quiet surface upon which you can repeatedly activate the tuning forks, with minimal sound intrusion from the activation itself.

Softly striking the tines of the High Ohm Tuning Forks together is another activation method. Tapping the metal tines together produces a much louder sound that will fill a large room, but can easily overwhelm a treatment room.

Once you have activated the High Ohm Tuning Forks sweep the energetic field around the body—and the environment of the room—to begin clearing, balancing, and strengthening the energy. As a general rule, hold the activated tuning forks a comfortable distance above the body, and use either a sweeping motion or a rolling technique. As you sweep or roll the tuning forks over the body, keep the tuning forks in a direction mostly parallel to the body. Point the activated tuning forks into the corners of a room to help disperse stuck energy. The High Frequency Ohm Tuning Forks are a lovely way to end a treatment. The higher octave(s) sounding in the energy field above and around the body helps to dissipate negative resonance and strengthen the Qi.

In this system, the healing intelligence of the body extends to the energetic body. By using Ohm and its overtones, the necessary sound intervals will be assimilated by the body's energy centers (chakras), facilitating a return to homeostasis and balance necessary for healing. As all color exists in white light, all sound exists in Ohm. Underlying this system is a belief in the body's healing intelligence: the body will take in the overtones and intervals it needs from the fundamental of Ohm.

Rhythm and the Sound Healer

There is an art to applying tuning forks to the body; it is a rhythmic process in which the practitioner is entraining his or herself to the rhythms of the other person. Attention to how slowly or quickly one approaches the body of the person being treated is essential. For an optimal experience, every movement should be considered. Be mindful of your body mechanics and the fluidity of movement as you move the tuning forks away from the body for re-activation, and then return them to the body—this is a rhythm. The pace should be unhurried.

A Musical Rest

It is a natural biorhythm for the body to be active—and to rest. The same concept applies as well in a treatment setting. The body does not need a constant vibration or sound applied to a muscle or acupuncture point in order to be effective. The natural decay of the sound of the tuning forks allows for a time of silence. This silence allows the body to absorb what it needs. In music this silence or pause is called a rest.

The musical rest that naturally occurs (still point) during a treatment allows the body to assimilate the sound medicine offered. This rest exists each time the practitioner pauses to re-activate the tuning forks, and at the end of a treatment. In addition, there is a natural sound "decay" characteristic of acoustic instruments which contributes a wealth of healing harmonics and overtones. These subtleties as well as the rhythmic relationship between the practitioner and the person being treated are important aspects of a Sound Healing treatment.

In a treatment setting, when vibrational sound is being directed on the body, it is of key importance to provide these musical rests or still points so the body can assimilate the applied tone and its subsequent overtones. The rest between applications allows the sound to travel to where it is most needed in the body. In Cranial Sacral Therapy, when the still point is accessed, the cerebral spinal fluid returns to a natural and healing rhythm. When this occurs, the breath deepens and releases, and the body regroups and returns to homeostasis. It is here, within the stillness that we replenish and regenerate.

Again, the pause in between applications is a critical musical rest that allows the sound vibration to travel where needed. This rest or silence allows the body to take in the vibration and manifest results. A sign of a confident Sound Healer is one who is able to attune to the rhythms, and who knows when to step back and allow the still points to occur.

How to care for Tuning Forks

To clean, wipe the tuning fork with a soft cloth, particularly the bottom of the stem, moistened with a dab of tea tree oil as a non-toxic disinfectant.

Please Note: the weighted ends of the Mid, Low and Osteo Ohm Tuning Forks should not be removed or adjusted as this will change their precision tuning.

Ohm Therapeutics Tuning Forks

Ohm Therapeutics Tuning Forks are made in the United States with aerospace and medical grade aluminum alloy and are precision tuned to within 0.5% of their indicated frequency.

This system is, in part, about the healing connection between our selves and the Earth. In order to reduce environmental impact, we have chosen not to color our tuning forks. This eliminates the de-greasing process necessary before painting or anodizing, all of which produce toxins to the environment.

In addition, we have the bottom of the stem of our tuning forks sanded smooth to ensure maximum comfort when the Ohm Tuning Forks are applied to the body.

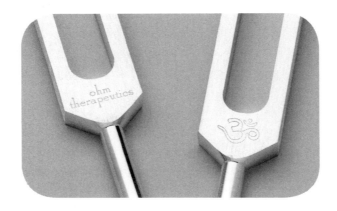

A QUICK LOOK AT TUNING FORKS

Tuning forks are commonly used in the medical profession. Nurses and doctors use tuning forks to test nerve conduction for problems such as hearing loss, and for neuropathies associated with conditions such as diabetes. Musicians use tuning forks to tune their instruments, and the police use tuning forks to test the frequency of their radars! Massage and Polarity Therapists, Spa Technicians, Reiki Practitioners, Acupressurists, Chiropractors, Energy Workers, among other health care practitioners, are incorporating Vibrational Sound Healing into their practices with tuning forks.

Acupuncturists find that tuning forks are an excellent addition to their practice, utilizing the benefits of sound vibration and the body's meridian system. For needle sensitive patients, the non-invasive nature of a tuning fork is an effective alternative.

Egyptologist and writer John West discovered a granite obelisk in Egypt dating from approximately 1470 BC. It is shaped like a giant tuning fork and when struck with the hand, resonates like a tuning fork. Some speculate this type of instrument helped move the stone blocks to build the pyramids.

In 1711, John Shore, an English trumpet player, is said to have constructed the first modern tuning fork.

Even the pop culture icon of the '60s, Samantha of the TV show *Bewitched,* receives a treatment from Doctor Bombay, showing that the idea of therapeutically directing sound vibration with tuning forks is not a new idea at all.

PART V

Essential Tools

Ohm Therapeutics is a vibrational healing system featuring *Ohm Tuning Forks + Music* composed in the same key or frequency. The music is designed to harmonize with the Ohm Tuning Forks and can lengthen or extend the sound vibration, enhancing your overall treatment. The tuning forks draw the music into your physical body, deeply attuning it to the Ohm vibration. The following are descriptions of the tools needed for self-treatment and treating others.

Basic Practitioner Tools
- 2 Mid Ohm Tuning Forks
- 1 Low Ohm Tuning Fork
- 1 Practitioner Activator
- CD *There's No Place Like Ohm Vols. 1 or 2*

Highly Recommended Tools
- High Frequency Ohm Tuning Fork Set
- Sonic Ohm Tuning Fork
- Additional Low Ohm Tuning Fork
- Osteo Ohm Tuning Fork
- Additional Practitioner Activator
- CDs: *In the Key of Earth*
  Vibrational Healing Music (Available March 2009)

Two Mid Ohm Tuning Forks

The most frequently used Ohm Tuning Fork is the Mid Ohm, which is 136.1 hz. Two Mid Ohm Tuning Forks are recommended to work bilaterally and distally on the body, and to double the healing resonance. Together, two Mid Ohm Tuning Forks form an Ohm unison. Because the body is symmetrical (left and right, front and back) it is important to work in a balanced manner.

Low Ohm Tuning Fork

The Low Ohm Tuning Fork is one octave lower than the Mid Ohm, meaning it is the same tone at a lower pitch. This translates to half the frequency of the Mid Ohm Tuning Fork, or 68.05 hz.

The Low Ohm Tuning Fork has a deep and earthy reverberating quality with lovely overtones when activated, and is an excellent choice for working on the body, especially the lower body, as it is particularly grounding. Using two Low Ohm Tuning Forks (Low Ohm unison) together is rooting. When doubled, the deep and resonant quality of the Low Ohm Tuning Forks facilitates loosening deeply stuck energy.

Ohm Octave ~ A Musical Interval

Activating the Low Ohm Tuning Fork in combination with the Mid Ohm Tuning Fork creates an Ohm octave. Octaves are potent, beautiful and healing musical intervals: they move and strengthen Qi as well as clear energetic blockages.

Osteo Ohm Tuning Fork

Recent research indicates that low sound frequency enhances bone density in animals and humans. The results show that frequencies which lie between 25-150 hz have been instrumental in healing muscles, bones, tendons, and ligaments, and helpful in relieving pain. In fact, researchers found that application of vibrational frequencies in this range helped to relieve pain in over 80 % of the participants with both acute and chronic pain. Ohm Therapeutics includes three Ohm frequencies within this range: the Mid Ohm at 136. 1 hz, the Low Ohm at 68.05 hz, and the Osteo Ohm at 34 hz.

An exciting part of this research comes from studying the vibration of purring cats. Dr. Clinton Rubin researched and discovered that frequencies between 25—50 hz can increase bone density by as much as 20 %. The application of this frequency can heal broken bones and encourages bone regeneration. These discoveries are also reminiscent of the predictions by Edgar Cayce, the Sleeping Prophet, who in the 1930's said that by 2020 we would be able to regenerate limbs using sound. For those of us who have been employing Sound Healing in our treatments for years, this news is exciting but not surprising!

The Osteo Ohm in the Ohm Therapeutics healing system is 34 hz and falls into the vibrational range of Dr. Rubin's cat purr research, 25-50 hz. The Osteo Ohm Tuning Fork is one octave lower than the Low Ohm Tuning Fork at 68.05 hz and two octaves lower than the Mid Ohm Tuning Fork at 136.1 hz.

Use the Osteo Ohm Tuning Fork on the vertebrae of the spine and on the joints of the body to relieve pain, reduce inflammation and increase bone density.

Low Ohm Octave

Activating the Low Ohm Tuning Fork in combination with the Osteo Ohm Tuning Fork creates a Low Ohm Octave. The Low Ohm Octave combines the therapeutic attributes of the lower frequencies along with the cathartic movement created by the musical interval of an octave. The Low Ohm Octave is particularly helpful with joint and bone pain. For example, many have reported feeling relief from knee pain with the lower Ohm frequencies.

Practitioner Activator

Designed for use with all tuning forks, the Practitioner Activator is comfortably worn around the leg. The strap is adjustable and the thickness and density of the rubber surface chosen to minimize the sound of the "hit" from activating the tuning forks.

High Ohm Octave Tuning Forks

The High Ohm Tuning Forks are ideal for use in the energetic field around the body. The sound wave created by the High Ohm Tuning Forks is joyful, bright, energizing, soothing and therapeutic; it restores balance to the energetic field around your body, thereby benefiting your physical body. Activating the High Ohm Tuning Forks at the end of a treatment dispels negative resonance and creates clarity, facilitating the transition from a deeply relaxed to a more wakeful state.

The High Ohm Tuning Forks are respectively one and two octaves higher than the Mid Ohm Tuning Forks, which means they are the same tone at higher pitches. This means that the larger of the two tuning forks in this set is 272.2 hz, which is twice the frequency of the Mid Ohm Tuning Fork, and the smaller is 544.40 hz, which is four times the frequency of the Mid Ohm Tuning Fork. The two high frequency forks are most effective when activated together, to form the High Ohm Octave.

High Ohm Octave (272.2 hz + 544.4 hz)   Sonic Ohm Octave (544.4 hz + 1088.8 hz)

The Sonic Ohm Octave in the highest octave in the Ohm Therapeutics system. This octave occurs when the Sonic Ohm Tuning Fork (1088.8 hz) is activated with the smaller tuning fork in the High Ohm Octave Set (544.4 hz).

SPACE CLEARING: Energetically a room, like a body, has a center and is responsive to the movement of Qi or the *Feng Shui* of sound. The High Ohm Tuning Forks are excellent tools for clearing and resetting the energy of a room. It is beneficial to clear and balance the space before and after a treatment.

Sonic Ohm Tuning Fork

The axiom "As Above, So Below" is beautifully expressed in this spectrum of Ohm Tuning Forks, with the Sonic Ohm being the highest frequency (with the Osteo Ohm being the lowest) in this system. The Sonic Ohm measures 1088.8 hz, which is an octave higher than the 544.4 hz tuning fork. The Sonic Ohm has a crystalline clarity and combines beautifully with each of the Ohm Tuning Forks. This higher expression of Ohm is also an earth tone, and has centering and grounding properties.

The Sonic Ohm Tuning Fork is an excellent sound tool for balancing the energetic fields or chakras of your body, and can be effective for clearing crystals.

CHAKRA BALANCING: The High Ohm Octave and Sonic Ohm Tuning Fork are effective tools to direct sound vibration in a room and in the body's energetic field. Chakra balancing is one such practice. The High Frequency Ohm Tuning Forks can cut through and clear stuck or stagnant energy, helping to restore balance to the body's chakras or energy centers.

CRYSTAL CLEARING: Crystals are subject to many of the same environmental influences and stresses that people experience. A crystal also benefits from clearing and rebalancing its physical body and the energetic field surrounding it. The high frequency tuning forks are effective tools because they resonate well with the frequencies characteristic of crystals.

Note: Activating or touching the tuning fork on the crystal is not recommended. The sound vibration will travel easily and resonate with the crystal by holding in close proximity, without physical contact.

Because crystals grow beneath the Earths' surface, where they remain until they are unearthed, the Ohm vibration is a natural choice for resonating with and resetting the intrinsic earth energy of a crystal.

> *"Science has now established the fact, long recognized by psychics and our ancestors, that the atoms in plants, crystals and the human body are tiny harmonic resonators in a constant state of vibration."*
> ~ ANDREW GLADZEWSKI, SCIENTIST

CDs ~ Music in the Key of Ohm

There's No Place Like Ohm Vols. 1 & 2 (Lemniscate Music, 2002 & 2005), *In the Key of Earth* (Sounds True, 2007) and *Vibrational Healing Music* (Sounds True, Spring 2009) are ideal for stress reduction, pain management, meditation and yoga.

> *"The time is fast approaching when humans will select their music with the same intelligent care and knowledge they now use to select their food. When that time comes, music will become a principal source of healing for many individual and social ills, and human evolution will be tremendously accelerated."*
>
> ~ Corinne Heline, Author and Mystic (1882-1975)

Tuning Forks + Music

At the center of this system are *Music + Tuning Forks* tuned to the same frequency of Ohm. Each of the aforementioned CDs are musical works in the tonic of Ohm that feature the rhythms and pulses of the natural world. Listening to music designed to resonate with the Ohm Tuning Forks creates another healing dimension. The wave, rhythm and tonal properties of these compositions are intentionally designed for use with the Ohm Tuning Forks. Our bodies respond to the biorhythms present in the music, which are a reflection of the cycles and seasons that surround us. Many find this orbital tone of Ohm to be intrinsically soothing and energizing.

Underlying the music of each CD featured in this system is a wave of sound that carries you on a deeply relaxing voyage. This sound wave has a gentle undulating quality; the levels were carefully modulated to rise and fall like an ocean wave. The effect is transporting and rich with overtones.

The heart of this medicine is experienced in several ways: with the Ohm Tuning Forks and music, through listening, and through skin and bone conduction.

PART VI

Therapeutic Application

"Because the power of harmony evokes moods and feelings, harmony should be kept simple when working with physical disease."
~ Kay Gardner, Sound Healing Pioneer & Musician (1941-2002)

Central to Ohm Therapeutics Sound Healing is the belief in the body's natural healing intelligence. The body's innate healing abilities are facilitated through a "sound environment" created by music and tuning forks tuned to the fundamental tone of Ohm and its octaves.

Because Ohm is an earth tone, it is characterized by pulses, cycles and seasons. Consequently, resonating with Ohm re-connects us to these fundamental rhythms of life. When we experience Ohm, through chanting, toning, listening to music in the key of Ohm and/or with sound healing tools, we entrain with natural cycles. The result is balancing and revitalizing.

Anatomical muscle and acupressure locations are described in the Self Treatment and Treating Others section of the manual and DVD. Specific acupuncture points are identified with each treatment protocol. Applying tuning forks on acu-points is an effective way to utilize the meridian system and energetic pathways of the body. For those working from different bodywork perspectives, some muscle locations are provided to further assist in identifying key treatment placements.

The following graph demonstrates the results of a technique called *Electro Meridian Imaging*, developed by Dr. John Amaro, and measured by Dr. Jimmie McClure, Holistic Chiropractor and Meridian Therapy Practitioner.

This graph represents a direct visual image of frequency. The numbers in the left column indicate frequency. The highlighted bars represent specific acupuncture points. The numbers above the bars indicate the measured frequency of the acupuncture points.

This graph simply demonstrates and affirms the presence of frequency in the acupuncture meridians and points of the body.

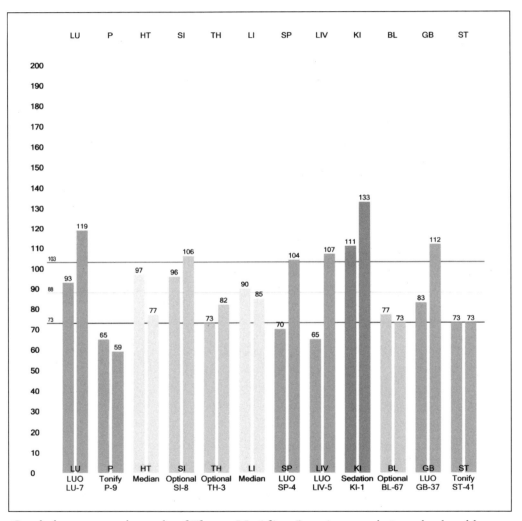

Graph demonstrates the results of **Electro Meridian Imaging**, *a technique developed by Dr. John Amaro.*

Acupuncture, Trigger & Reflex Points

Acupuncture is an ancient discipline dating back thousands of years. It began with touch on points (acupressure) first using fingers and then rocks, evolving to needles, and now incorporates the use of tuning forks. It is one of the oldest medicinal practices in the world and its healing benefits have been well documented over time. The acupuncture meridians are associated with the body's organs. The application of tuning forks to the corresponding acupuncture points allows us to access these vital energy centers. At the center of the body's energetic matrix is the Microcosmic Orbit which travels along the midline of the front and the back of the body.

Microcosmic Orbit

The Microcosmic Orbit is an alchemical mix of yin energy, believed to be fed by the Earth, and yang energy, believed to be fed by the heavens. Simply put, the body's meridians are fed by the energy of the Microcosmic Orbit.

The Microcosmic Orbit of the body can be likened to the electron's orbital path around the nucleus of an atom, as well as the Earth's orbital path around the Sun.

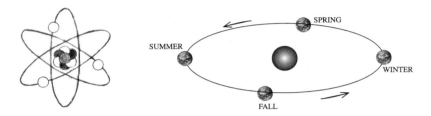

Tuning forks are a natural fit with acu-points. Stimulating these points (activating the Qi) by directing sound vibration through the use of tuning forks is extremely effective. In addition, the tuning forks fit nicely on acu-points. The body is made of vibration and responds well to vibration, and research shows that cells communicate with one another via vibration. By introducing vibratory sound via acupuncture points along the body's energetic pathways, we have the opportunity to attune the body on a cellular level.

Trigger Points

Trigger Points are points on the muscle that have become blocked by strain, injury, stress or poor posture. Some trigger points and acupuncture points are the same. Stimulation of these points assists in releasing the blocked energy.

Tuning forks are effective in treating trigger points. They save wear and tear on your hands, thumbs and fingers by vibrating the point. The sound vibration and pressure applied from the stem of the tuning fork help to dissipate blocked energy, and consequently relieve pain and bring Qi back into the area.

Reflexology

Reflexology is an ancient healing art based on the belief that reflex points exist in the feet, hands and ears and correspond to all parts of the body. For example, the whole body can be treated reflexively by using tuning forks on the indicated placements on the foot. In addition, stimulating these points with tuning forks helps improve circulation and enhances the flow of energy. Working reflexively is an excellent method for those who are unable to tolerate direct tuning fork application on a particular area of the body.

> *To Refresh...*
>
> ## OHM THERAPEUTICS GUIDING PRINCIPLES
>
> 1. A belief in the body's natural healing intelligence.
>
> 2. A knowledge that we are made of vibration.
>
> 3. A belief that applied vibration with Ohm Tuning Forks helps to remove blockages in the body's energetic pathways.
>
> 4. An understanding that an individual's biorhythms and cycles are intimately connected to those of the Earth.
>
> 5. A belief that sympathetic resonance with the Earth helps to restore balance, establishes homeostasis, and promotes healing.

Ohm is well described as an earth tone based on our planet's orbital path around the Sun. The Ohm vibration facilitates homeostasis, where profound healing begins. For this reason, working with the Ohm Tuning Forks and music composed in the same "key" or frequency of Ohm is a perfect complement to bodywork and energetic therapies. The beauty of this system is its simplicity and effectiveness.

Practice Application

Begin practicing your basic technique with two Mid Ohm Tuning Forks. In each hand, hold the stem of the tuning fork with a firm yet relaxed grasp between your thumb and fingers. Relax your hands, wrists and arms, and take a deep breath. Strike the grooved edge of the weighted end of each fork at an angle on the Practitioner Activator. If you hit the flat part of the weight of the tuning fork on the activator it can be noisy and will not provide enough kinetic momentum. Make sure you activate the tuning forks with enough energy to provide a strong and lasting vibration. Then, in concert, bring the activated tuning forks to each ear for listening; a comfortable distance is between 4" and 6" from each ear. This produces a "stereo effect" which is greatly enhanced by playing the recommended musical CDs.

Listening begins the sound healing process. A well activated tuning fork will have a clear strong tone that will sustain for about 20 to 30 seconds. Remember to breathe deeply and relax, as you repeat the process of tuning fork activation. By repeating this process and listening to the activated tuning forks, the movement becomes smoother and more fluent. Practice this technique before applying the tuning forks to your own or another's body.

Human Touch

When applying the tuning forks to your body, practice fanning out the last two to three digits of each hand. This way, when the bottom of the stem of the tuning fork makes contact with the body, so does that part of your hand. The warmth and familiarity of human touch is essential. Gently palpate or touch the area where you will place the activated tuning fork. Allow your fingers and hands to rest around the area where the tuning fork makes contact with the body. Touch communicates to the body, and prepares it to receive the applied sound vibration. This helps prevent tension and supports the tuning forks, and will be reassuring to the person who is receiving the treatment. See DVD.

KEY POINTS

1. Ohm Therapeutics Vibrational Healing with Tuning Forks is based on the understanding that musical tones assist healing.

2. Sympathetic Resonance with the earth tone of Ohm experienced physically and auditorily, promotes a positive flow of energy in and around the body.

3. Sound vibration opens the energetic pathways where the Qi or natural life force flows. As a result, energy blocks are removed, increasing the flow of Qi, and facilitating homeostasis where profound healing begins.

4. At the center of this healing system are *Music + Tuning Forks* in the same beneficial frequency of Ohm.

5. The body is able to assimilate Ohm and all its beautiful overtones and utilize these healing properties on a vibratory level.

Holding Sacred Space

One of the most important things to establish
in a healing treatment is sacred space.

Sacred space creates room for knowledge, intuition and intention to unfold.

A treatment space is like the Medicine Wheel. The sacred directions
and elements surround and assist you at center. The center is empty and
holds the silence. From the silence comes forth the vibration.

Notice your breath and the breath of the person being treated.
It establishes the rhythm of the treatment.

Use the tuning forks as sacred healing tools to clear and charge the space.
Your feet are planted firmly on the ground, establishing contact with the
Earth; your head is clear and receives the sky energy.

Hold the sacred space and allow for the wisdom of the vibration to come forth.

As you facilitate the Sound Healing treatment,
the natural intelligence of the body unfolds.

This is a quantum moment in time,
a microcosm of the macrocosm.

PART
VII

Treating Self & Others

"Home is where we start from..."
~ T.S. Elliot, Poet (1888-1965)

PRACTICE ON YOUR OWN BODY BEFORE PRACTICING ON OTHERS. This is important for two reasons:

1) Practice allows you to become familiar with activating the tuning forks and the amount of pressure used in applying them to the body.

2) Once you understand how it feels to have tuning forks placed on your own body, you will have a greater sensitivity when treating others.

Begin by applying the activated tuning forks on your body, for example, below your elbow on the fleshy part on the top of your forearm, or on the top of the shoulder near the neck. As you feel comfortable, move to other areas of the body.

Pay attention to how the sound vibration feels on your body. Can you feel tight muscles begin to relax? Can you feel the resonance in your bones? Observe how the vibration sustains or how quickly it dampens. The sustain lasts longer when the tuning fork is well activated, and when it is held and placed properly.

The Ohm Tuning Forks are very effective for use in meditation as they help block external noise and distraction, bringing in balance and stillness. The Ohm Tuning Forks and the Ohm Singing and Crystal Bowls create a beautiful salutary tone to begin and end a yoga session. Many people use the Ohm Tuning Forks to take a "sound break" in the middle of a busy day. Listening to Ohm is calming and energizing.

Self Treatment

The following five examples demonstrate where to place the Ohm Tuning Forks on your body, and describe the associated benefit. Your body position should be comfortable and relaxed while practicing the following tuning fork applications.

TREATMENT #1   PALM OF HAND

Location: Palm of hand

Application: SINGLE

Apply Mid Ohm Tuning Fork to the center of the palm of your hand.

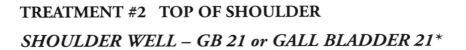

Direction: Straight down on your palm, almost perpendicular to your hand.

Energetic/Benefit: Releases tightness and tension from neck, shoulders, chest, arm, and hand. Relaxes the body and calms the spirit. Helps one to feel more connected to the self and to the world.

TREATMENT #2   TOP OF SHOULDER
*SHOULDER WELL – GB 21 or GALL BLADDER 21**

Location: Top of the shoulder, near the neck

Application: SINGLE

Apply Mid Ohm Tuning Fork to the top of the shoulder, in the shoulder well, about 1/3 the distance from the side of the neck and the end of the shoulder.

Direction: The stem of the tuning fork is angled toward the neck; the weighted end is angled slightly outward.

Energetic/Benefit: Releases neck, shoulder and head tension. Helps relieve stress, headaches, and facilitates deeper breathing. This point relaxes and opens the entire upper body. The shoulders drop, the head clears and the breathing becomes deeper.

*Contraindicated for pregnancy

TREATMENT #3   BOTTOM OF FOOT
BUBBLING SPRING – KI 1 or KIDNEY 1

Location: Bottom of foot

Application: SINGLE AND DOUBLE

Gently curl the toes, and find the point in the depression in the middle of the bottom of the foot, and apply Mid Ohm Tuning Fork just below the ball. For Double Application, use 2 Mid Ohm Tuning Forks, the Ohm Octave or the Low Ohm Octave.

Direction: The direction of the tuning fork(s) is angled toward the ball of the foot.

Energetic/Benefit: Grounding. Gently draws tension from your head, shoulders and upper body. Connects you to the Earth and brings energy to your feet and legs.

TREATMENT #4   STERNUM
PRIMORDIAL CHILD – REN 17

Location: Middle of sternum or chest

Application: SINGLE

Midway between top and bottom of sternum, between the breasts, approximately level with the 4th intercostal space. Apply Mid Ohm Tuning Fork to the small notch at this point, in the middle of your sternum.

Direction: The position of the tuning fork is straight down on the body, almost perpendicular to the sternum.

Energetic/Benefit: Soothing, energizing and centering. Initiates deep breathing and relaxation by accessing the innocence of the heart. Helps to heal hurts and tap into unconditional love.

TREATMENT #5 STERNUM + LOW ABDOMEN
*REN 17 and REN 6**

Location: Middle of sternum or chest and approximately 2 finger widths below the navel.

Application: DISTAL (Ohm Octave)

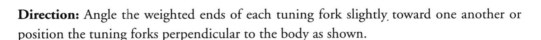

Apply the Mid Ohm Tuning Fork on sternum (middle of chest) and the Low Ohm Tuning Fork 2 to 3 finger widths below the navel.

Direction: Angle the weighted ends of each tuning fork slightly toward one another or position the tuning forks perpendicular to the body as shown.

Energetic/Benefit: Connects the lower abdomen with the heart, the adult and the child within. Helps with insomnia, restlessness and anxiety as it deeply relaxes the entire body. The low abdomen or "Sea of Qi" is believed to be the storehouse of heavenly yang and earthly yin energy.

*Some sources contraindicate for pregnancy.

Treating Others

After practicing the basic techniques presented for correctly holding, activating and applying the tuning forks to your own body, you will feel more comfortable treating others. The following section will address the very important topic of body mechanics, and present application methods and treatment techniques for both the physical and energetic body.

The treatment protocols presented are successfully used by a diverse array of healthcare practitioners. The efficacy of the various point and muscle locations has been clinically demonstrated for years by Acupuncturists, Massage Therapists, Chiropractors, and others working in the healing arts.

Refer to the following charts summarizing the five step application process and the four application methods.

MID, LOW & OSTEO OHM TUNING FORKS

APPLICATION PROCESS (on the body)

1. **FIND** the location for desired application.
2. **PALPATE** the location.
3. **EYE** the location.
4. **ACTIVATE** the tuning forks, while eyeing the location.
5. **PLACE** the tuning forks on the body.

APPLICATION METHODS (on the body)

1. **SINGLE APPLICATION:** one tuning fork.
2. **BILATERAL APPLICATION:** two tuning forks; apply one each on bilateral points or body locations.
3. **DISTAL APPLICATION:** two tuning forks; apply on distal areas of the body, in order to energize the area between the points or body location.
4. **DOUBLE APPLICATION:** two tuning forks; apply both to one point or body location.

Body Mechanics

The position of your body when applying the tuning forks is extremely important. Good body mechanics will aid you in every aspect of your treatment, from the activation of the tuning forks to the applications suggested in the following treatment protocols.

The possibilities of straining (e.g., your back and shoulders) greatly increase with poor body mechanics. Pay careful attention to your body position as you work around the treatment table. Be mindful of your posture and movement as you reach to locate and place the activated tuning forks. Please refer to the accompanying DVD to help you determine your body position appropriate to each treatment.

When you begin the treatment make sure the recipient is in a comfortable position. During the treatment, check in each time the recipient changes positions to make sure he or she is still comfortable. In any treatment setting, there may be moments when a tight muscle or an injured area that is being worked on could be uncomfortable. Explain this to your client, and ask them to let you know if you need to lighten the pressure of application, or cease treating in that area. The rhythm and resonance that develops between you and the recipient of the treatment will eventually develop and flow, guiding and informing how you work together. Communication and good body mechanics for both of you are always of key importance.

The Spinal Treatment *(#2 Hua Tuo Jia Ji Points)* and Upper Chest *(#10 Central Treasury—Lu 1)* are two examples that illustrate some of the finer points of body mechanics when using tuning forks. While watching the DVD, take special note of the practitioner's body position.

The Spinal Treatment starts at the neck and ends at the sacrum. A good way to begin the Spinal Treatment is to stand/position your body at the head of the table. Activate the tuning forks and slowly bring them in toward the recipient's ears to begin the body attunement through listening. Next, treat the indicated points along the cervical region of the spine. Your body mechanics from this point on will depend on your and the recipient's size, and the height of your treatment table. If, from this position, your reach is too far to comfortably treat the lumbar spine, then move from the head to the side of the table.

When treating Lung 1, apply the tuning forks to the indicated points bilaterally. Position your body position at the head of the table. In order to follow up with the next application—double on Lu 1—you will need to move to the right or left side of the table. From this position you can easily and without strain reactivate the tuning forks for the double application. As the applied sound vibration travels, a natural musical rest will occur as you once again reposition yourself on the opposite side of the treatment table. Now repeat the double application to the other upper chest point. If you find yourself reaching over to the other side of the body, recheck your body position and adjust.

Practice your technique and body mechanics to determine how you will to work in a treatment setting. Good leverage, the direction of the tuning fork on the body and the ability to use your body position will assist as you apply the tuning forks. For these reasons, you may want to experiment with your treatment table in a slightly lower position than you are normally accustomed to. You will know what works best for you.

HIGH FREQUENCY TUNING FORKS

APPLICATION PROCESS (around the body)

1. **ACTIVATE** the tuning forks in a rhythmic manner, first one and then the other, and gently enter the energetic field around the body with the intention of clearing away negative resonance, and strengthening the energetic field.

2. **DIRECT** the tuning forks above and around the body, keeping them somewhat parallel to the body. You can also direct the forks toward the corners of a room to move and dispel stuck energy.

APPLICATION METHODS * (around the body)
use High Ohm Octave or Sonic Ohm Octave

1. **ROLLING:** Visualize the motion of a wheel rolling as your hands, each one holding an activated tuning fork, roll over one another. Begin with a forward motion, rolling away from and then back toward your body. The speed at which you 'roll' or move the activated forks will be a matter of comfort. As you practice, you will notice that at times you will roll them more widely apart, and at other times more closely together. A wider roll is generally more dispersing, whereas a tighter roll can create a more focused sound vibration effective for breaking up an energy block.

2. **CADUCEUS WEAVING:** Trace a figure eight motion above the spine to reinforce this subtle energy. Use this motion both up and down the area above the spine.

The Caduceus or the "Wand of Hermes" is usually depicted as a herald's staff entwined by two serpents in the form of a double helix. At the top of the staff there are often wings. The caduceus is the modern symbol of Western medicine, some say with ancient roots in astrology and the god Chiron.

The rod of the Caduceus also represents our body and the wings our shoulders. Overall, this powerful image symbolizes humanity's ability to connect to spirit. Kundalini energy is also depicted by the spiraling movement of the intertwining snakes. This energetic force awakens us to a greater understanding of our psycho-spiritual selves.

PART VIII

Treatments

"Accessing the body's meridians through points or gateways connects the pulsing eco systems of the body in the same way that rivers and streams fluidly connect the eco systems of the Earth."

THE FOLLOWING TREATMENT DESCRIPTIONS AND ILLUSTRATIONS are intended to help guide you as you learn to incorporate sound healing with tuning forks into your existing practice. The following twenty treatment protocols demonstrate how to use tuning forks on specific acupuncture, trigger, and reflex points. Each of these treatment protocols is demonstrated on the companion DVD. Refer to the DVD to help reinforce correct body mechanics and the finer points of tuning fork application. In sum, the treatments presented work as a holistic body treatment and/or the individual protocols can be selectively integrated into your practice.

The Spinal Treatment is an extremely beneficial treatment for the whole body and is highly recommended to introduce into any healing session. You can access deep healing and enervate the entire body through the para-vertebral or *Hua Tuo Jia Ji* points. These points found beside the spine are gateways for accessing the Microcosmic Orbit and energetic pathways of the body.

The spine is our *axis mundi*, with the nerve centers that correspond to all of the body's vital processes. Applying the Ohm Tuning Forks distally along these nerve plexus is an excellent way to energize the entire organ system, nourishing the Qi and attuning the overall body.

These treatments were designed to help guide you on your journey to Sound Healing using Ohm Tuning Forks. Relax and Enjoy!

TREATMENT # 1   LISTEN

Where: Ears

Body Position: Prone or Supine

Application: BILATERAL

Place 2 activated Mid Ohm Tuning Forks about 3-5 inches from the recipient's ears. Ask if they can hear the sound comfortably.

Direction: The weighted ends of each activated tuning fork are held near the ears.

Energetic/Benefit: The act of listening to the tuning forks begins the sound healing journey. The extraneous noise of the world begins to dim as the Ohm tone brings focus to the moment. The breathing deepens and the body begins to relax.

TREATMENT # 2 SPINAL TREATMENT
HUA TUO JIA JI POINTS

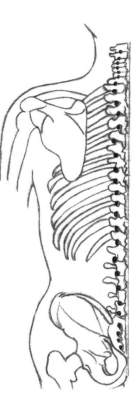

Where: Back

Body Position: Prone

Location: These points are located right next to the spine, between the transverse process of each vertebra, from the cervical through the sacrum. These points are known as the **Hua Tuo Jia Ji** points *(pronounced Wah Too Oh Gee Ah Gee)* in Chinese Medicine.

Application: BILATERAL

Apply 2 activated Mid Ohm Tuning Forks bilaterally, starting with the cervical spine, at the base of the skull, moving down through the thoracic and the lumbar spine, and concluding with the sacral foramen. The 8 sacral foramen (openings) are also called the 8 **Ba Liao** *(pronounced Bah Lee Ah Oh)* and consist of 4 points each on either side of the midline of the sacrum.

Ask the recipient if they are comfortable touching their tongue to the upper palate of their mouth, behind the front teeth, during this treatment. This enables the activated Qi to circulate the entire Microcosmic Orbit, where Earth and Sky energy mix to energize our bodies.

Direction: Angle the stem of the tuning forks slightly toward the spine.

Energetic/Benefit: The Spinal Points are an essential therapeutic treatment for the entire body. They restore Qi and blood flow to all of the organ systems and thereby tonify and energize the entire body. They balance and relax the nervous system. Collectively these points are located along the Microcosmic Orbit of the body. The bones are watery and hollow making them resonant chambers. The resonance and vibration created along the spine relaxes the musculature of the back and helps to adjust the vertebrae. The chakras are located at energetic centers along the Microcosmic Orbit and also benefit energetically from the Spinal Treatment.

TREATMENT # 3   BACK OF HEAD
HEAVENLY PILLAR – UB 10 or URINARY BLADDER 10

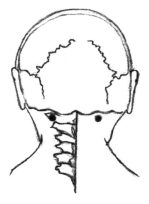

Where: Back of head

Body Position: Prone

Location: Just below the occipital protuberance on the back of the head, slightly above the hairline and on the lateral edge of the trapezius muscle, and about $1^{1/3}$ inches lateral to the spine.

Application: BILATERAL

Apply 2 activated Mid Ohm Tuning Forks bilaterally.

Direction: Angle the stem of each tuning fork toward the cervical spine or center.

Energetic/Benefit: These points open the gateway of energy flow between the head and the body. Activating these points helps to clears the head, relieve headache and neck rigidity, and improve concentration and vision. Some believe activating these points connects us to universal energy, which helps lessen fear and depression.

TREATMENT # 4   LOWER BACK

SACRUM (Sacred Bone)

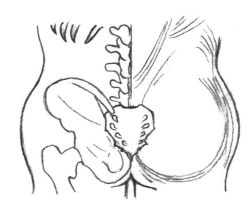

Where: Lower back

Body Position: Prone

Location: The large somewhat triangular shaped bone made up of the five fused vertebrae at the end of the spinal column, below the lumbar region and above the coccyx.

Application: DOUBLE

Apply the activated Ohm Octave to the center of the sacrum to open up the energy. Gently move the tuning forks back and forth around the edges of the sacral bone to create movement in the sacrum. Repeat with Low Ohm Octave.

Direction: In Double Application, apply both tuning forks to the same point. The weighted ends of each tuning fork are angled slightly outward.

Energetic/Benefit: The sacrum is a strong and protective bone and is the part of the pelvis that connects the torso to the legs. It is of utmost importance that the energy flows through the sacral area—from the torso to the legs and from the legs to the torso. The sacrum is the foundation of the spine and is a connecting point between the lower and upper parts of the body. Energy often gets stuck in the sacrum and can interfere with the flow of the Microcosmic Orbit. The Microcosmic Orbit is the central Qi system of the body and is fed from the energies of the Earth and Sky.

Applying the tuning forks to the sacrum with the powerful interval of the Ohm Octave helps to loosen stuck or rigid energy, free the hip area, and open the entire body stature.

TREATMENT # 5   BACK OF HEAD + LOWER BACK
HEAVENLY PILLAR – UB 10 and SACRUM

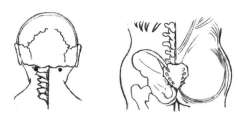

Where: Back of head and lower back

Body Position: Prone

Location: See Treatment Location #3 (UB 10) and #4 (Sacrum).

Application: DISTAL

Apply activated Low Ohm Tuning Fork to the center of the sacrum and an activated Mid Ohm Tuning Fork on the left side Heavenly Pillar (UB 10); repeat on the right side of Heavenly Pillar.

Direction: The weighted ends of the tuning forks are angled toward one another.

Energetic/Benefit: Creates resonance and energy between the head and the lower back, connecting the thinker with the doer, the cerebral with the physical. Applying the Ohm Octave to these locations helps to relieve headache and back pain, and creates a sense of wholeness, well-being and peace.

TREATMENT # 6 BETWEEN SHOULDER BLADES
RHOMBOID MUSCLES

Where: Region between shoulder blades

Body Position: Prone

Location: Area between thoracic vertebra and the insertion of the muscle into scapula.

Application: DOUBLE and DISTAL

Apply 2 activated Mid Ohm Tuning Forks to the rhomboids. Gently move tuning forks back and forth on the muscle.

Direction: In Double Application the tuning forks are both applied to the same location. The weighted ends of each tuning fork are angled slightly outward away from another. In Distal Application, position the tuning forks straight down on the muscle.

Energetic/Benefit: Releases tight shoulders and back, deepens breathing, and brings energy to the arms and hands. Opens and strengthens heart energy, helping to relieve stress, and encouraging sense of self and spontaneity.

TREATMENT # 7 SHOULDER BLADES
HEAVENLY GATHERING – SI 11 or SMALL INTESTINE 11

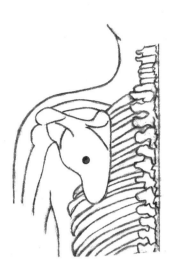

Where: Shoulder blades

Body Position: Prone

Location: On the scapula in the tender depression near the center, level with the 4th thoracic vertebra.

Application: BILATERAL and DOUBLE

Apply 2 activated Mid Ohm Tuning Forks bilaterally, and then use Double Application on each point.

Direction: In Bilateral Application the tuning forks are angled outward. In Double Application, apply both tuning forks to the same point. In each application, the weighted ends of the tuning forks are angled slightly outward.

Energetic/Benefit: Activating this point helps relieve pain in the shoulders and opens the energetic pathway through the body to the arms and hands. Assists in further opening and clearing the lungs, opening the chest area and creating an expansive and light feeling throughout the body. Helps to promote a connection to and a feeling of significance in the world.

TREATMENT # 8 BACK OF LEGS
HAMSTRING and CALF MUSCLES

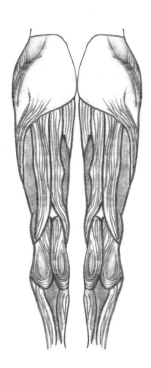

Where: Back of legs

Body Position: Prone

Location: Hamstring and calf muscles.

Application: DOUBLE and DISTAL

Apply 2 activated Mid Ohm Tuning Forks along the muscles (including trigger and anatomical points) using the tuning forks to gently rock the muscle back and forth.

Direction: In Double Application, apply both tuning forks to the same location. The weighted ends of each tuning fork are angled slightly outward. In Distal Application position the tuning forks straight down, almost perpendicular to the muscle.

Energetic/Benefit: Relaxes the legs and the lower back. Helps to release stress from the body, and strengthens stature.

TREATMENT # 9   BOTTOM OF FOOT
BUBBLING SPRING – KI 1 or KIDNEY 1

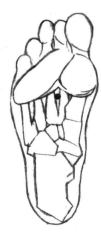

Where: Bottom of foot

Body Position: Prone or Supine

Location: Gently curl the toes, and find the point in the depression in the middle of the bottom of the foot, just below the ball. Be sure to remind the person receiving the treatment to relax their foot after you locate the point.

Application: BILATERAL and DOUBLE

Apply 2 activated Mid Ohm Tuning Forks to the indicated point. You can also use the Ohm Octave and the Low Ohm Octave on this point.

Direction: In Bilateral Application, apply tuning forks to both feet at the same time, and direct the tuning forks toward the ball of the foot. In Double Application, apply both tuning forks to the same point on one foot; repeat on the other foot for balance. The weighted ends of the tuning forks are angled slightly outward.

Energetic/Benefit: Grounding treatment which helps to gently draw excess stress from the head, shoulders and upper body. Activating this point energetically connects you to the Earth by bringing Qi to your feet and legs. Provides a sense of stability and security. Think of your body as a tree, with roots extending through your feet, powerfully connecting and tapping into the energy of the Earth.

TREATMENT # 10 UPPER CHEST
CENTRAL TREASURY – LU 1 or LUNG 1

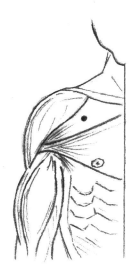

Where: Upper chest

Body Position: Supine

Location: Follow the lower border of the collarbone, and find the depression of the chest between the pectoral muscle and the shoulder bone. Come down approximately 1 inch to find the point.

Application: BILATERAL and DOUBLE

Apply 2 activated Mid Ohm Tuning Forks to the left and right points.

Direction: In Bilateral Application the stem of the each tuning fork is angled toward the muscle and center of the body. In Double Application, apply both tuning forks to the same indicated point.

Energetic/Benefit: Activating this point initiates a deep breath, and releases tension and stress, thus relaxing the chest and shoulders. The upper chest area houses our celestial Qi, replenishing life and affirming inner worth with every breath. Encourage the person receiving the treatment to release any anxiety and grief with each out breath. Deep breathing oxygenates our cells and connects us to the Universal Qi.

TREATMENT # 11 LOW ABDOMEN
*SEA OF QI – REN 6**

Where: Low abdomen

Body Position: Supine

Location: On the midline, 2 finger widths below the navel.

Application: DOUBLE

Apply the activated Ohm Octave to the indicated point.

Direction: The position of the tuning forks is straight down, almost perpendicular to the belly. In Double Application, apply both tuning forks to the same indicated point.

Energetic/Benefit: Activating this point harmonizes the Qi in the body by accessing and boosting the primordial Qi. Helps to relax the lower abdomen and the lower back and increases vitality. The Sea of Qi is believed to be the storehouse of heavenly yang and earthly yin energies. A very centering and grounding treatment. This point lies along the Microcosmic Orbit in the region of the second chakra.

*Some sources contraindicate for pregnancy

TREATMENT # 12 STERNUM
PRIMORDIAL CHILD – REN 17

Where: Sternum

Body Position: Supine

Location: Midway between the top and the bottom of the sternum, between the breasts, level with the 4th intercostal space. You will feel a small notch at this point.

Application: SINGLE

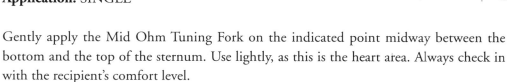

Gently apply the Mid Ohm Tuning Fork on the indicated point midway between the bottom and the top of the sternum. Use lightly, as this is the heart area. Always check in with the recipient's comfort level.

Direction: The position of the tuning fork is straight down, almost perpendicular to the body.

Energetic/Benefit: Soothing and calming. Initiates deep breathing and relaxation by accessing the innocence of the heart. Taps into unconditional love, helps to heal the hurts of life, and generates energy through strength of self. This point lies along the Microcosmic Orbit in the region of the heart chakra.

TREATMENT # 13   STERNUM + LOWER ABDOMEN
REN 17 and REN 6

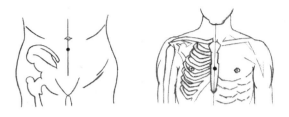

Where: Sternum and low abdomen

Body Position: Supine

Location: See Treatment Location #11 (Ren 6) and #12 (Ren 17).

Application: DISTAL

Apply the Mid Ohm Tuning Fork on sternum and Low Ohm Tuning Fork on lower abdomen.

Direction: Angle the weighted ends of each tuning fork slightly toward one another.

Energetic/Benefit: Connects the lower abdomen with the heart, the adult and the child within. Helps with insomnia, restlessness and anxiety as it deeply relaxes the entire body. A comforting treatment.

TREATMENT # 14   TOP OF SHOULDER
SHOULDER WELL – GB 21 or GALL BLADDER 21*

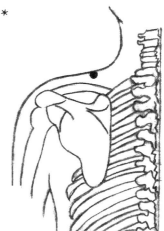

Where: Top of the shoulder, near the neck

Body Position: Supine

Location: The top of the shoulder, in the shoulder well, about 1/3 the distance from the side of the neck and the end of the shoulder.

Application: BILATERAL and DOUBLE

Apply Mid Ohm Tuning Forks bilaterally on each shoulder. In Double Application, apply both Mid Ohms on each point, first one shoulder and then the other.

Direction: In Bilateral Application the stem of each tuning fork is angled toward the neck. In Double Application the tuning forks are both applied to the same indicated point.

Energetic/Benefit: Releases neck, shoulder and arm tension and pain. Helps relieve stress, and facilitates deeper breathing. Activating this point relaxes and opens the entire upper body. The shoulders drop, the head clears and the breathing becomes deeper.

Contraindicated for pregnancy

TREATMENT # 15   JAW
MANDIBLE WHEEL – ST 6 or STOMACH 6

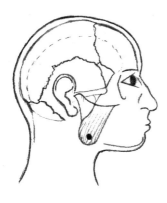

Where: Face

Body Position: Supine

Location: Muscle of the jaw. Approximately one inch diagonal from the corner of the jaw, on the edge of the masseter muscle.

Application: BILATERAL

Apply activated Mid Ohm Tuning Forks bilaterally to the indicated points on both sides of the jaw.

Direction: Angle the stem of each tuning fork against the masseter muscle, toward the back teeth.

Energetic/Benefit: Activating this point offers relief to those who clench and/or tighten their jaws, as it relaxes the entire jaw area. Relieves headache, stress and insomnia and encourages deeper breathing. Helps to release repressed emotions.

TREATMENT # 16 EYEBROW
GATHERING BAMBOO – UB 2 or URINARY BLADDER 2

Where: Eyes

Body Position: Supine

Location: In the hollow or bone notch at the medial end of the eyebrow, toward the bridge of the nose.

Application: BILATERAL

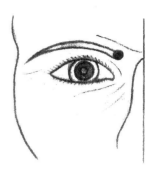

Apply activated Mid Ohm Tuning Forks bilaterally to the indicated points on each eyebrow.

Direction: The stem of the tuning fork is slightly angled up toward the inner eye. The indicated point is in the notch on the bone.

Energetic/Benefit: Activating this point helps to relieve over-thinking, anxiety and headaches, as well as eye strain and blurring. Quiets the mind, aids vision, and promotes a sense of inner peace.

TREATMENT # 17 EYE REGION

FRESH INNOCENT EYES – GB 1 or GALL BLADDER 1

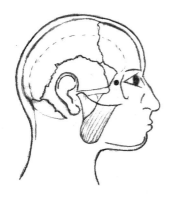

Where: Eyes

Body Position: Supine

Location: Approximately ½ inch from the outer corner of the eye, in the depression on the lateral side of the eye orbit.

Application: BILATERAL

Apply activated Mid Ohm Tuning Forks bilaterally on the indicated points.

Direction: Place the tuning fork stem slightly behind the eye orbit or bone and angle toward the eyeball.

Energetic/Benefit: Activating this point helps relieve headache and eye pain. Clears and brightens the vision.

TREATMENT # 18   TOP OF FOOT
STREAM DIVIDE – ST 41 or STOMACH 41

Where: Top of foot

Body Position: Supine

Location: On the top of the ankle, at the midpoint of the crease, in the depression between the two tendons of the muscle.

Application: BILATERAL or DOUBLE

Apply activated Mid Ohm Tuning Forks and/or the Ohm Octave.

Direction: In Bilateral Application angle the stem of the tuning forks toward the ankle crease. In Double Application, apply both tuning forks to the same point.

Energetic/Benefit: Helps to relieve ankle, foot, and leg pain. Application to this point clears the channel for the energy to connect with the Earth. Clears the mind, helps with headaches, and increases a sense of well being. Assists and supports the stance and the spirit.

TREATMENT # 19   TOP + BOTTOM OF FOOT
ST 41 and KI 1

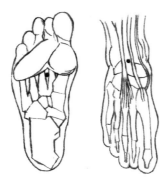

Where: Top and bottom of foot.

Body Position: Supine

Location: See Treatments and #18 (ST 41) and #9 (KI 1)

Application: DISTAL

Apply activated Mid Ohm Tuning Forks and/or the Ohm Octave. When using the Ohm Octave, apply the Mid Ohm Tuning Fork on ST 41 and the Low Ohm Tuning Fork on KI 1.

Direction: See Treatments #9 and #18.

Energetic/Benefit: Helps relieve foot, ankle, leg and lower back pain. Opens and disperses the energy, enabling the body to connect with the Earth Qi, and brings Earth energy up into the body. Energizing. Helps to restore a sense of joy.

TREATMENT # 20   ENERGETIC FIELD AROUND BODY
MICROCOSMIC ORBIT

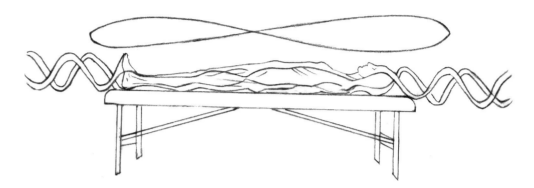

Where: Energetic field surrounding the body

Body Position: Prone or Supine

Location: In the energetic field around the body. Treatment Space.

Application: This protocol is recommended to end a treatment. With any combination of the High Frequency Ohm Tuning Forks, make a Figure 8 over the body with the sacrum at the intersection of the Figure 8. Use the Rolling Method over the body by moving from head to toe above the body. Use the Caduceus Weaving Method over the body.

Direction: Move tuning forks in a motion parallel to the body, as shown in the rolling and caduceus weaving techniques.

Energetic/Benefit: Adds a light vibration and balance to complete the treatment. Clears negative resonance released from the body. Strengthens the energetic field. Also clears a treatment space of stagnant or negative energy.

Ohm Singing Bowls

Ohm Singing Bowls add another harmonic dimension when activated before, during and after a Sound Healing Treatment. Activating a singing bowl can have a calming effect, and is a wonderful tone-setting salutation to begin and end a treatment. Imagine the rippling effect a stone has when it hits water, the rings enlarging as they move further from the center. The sound vibration of an activated singing bowl has a similar effect: the rings are like the emanating overtones.

The yarn mallet is the recommended tool for striking the bowl. It produces a beautiful gong-like effect. While rimming the outer edge of singing bowls creates a rich sound, it can easily overwhelm a small treatment room and have an over stimulating effect.

During the treatment, the tuning forks provide directed sound vibration to specific points and muscle. The singing bowls can also be activated at appropriate times during a treatment. They can be placed under the table or on a large enough treatment table. They can also be placed on the body or activated above the body. See DVD.

Ohm Singing Bowls are musically rich with healing overtones and are very effective and easy to play. Striking the side of a singing bowl with a suitable mallet produces a lovely tone that will continue to ring for up to a minute or even longer while the bowl is moved over and around the body, filling the room with sound healing tones and overtones. This effect is greatly enhanced when music in the key of Ohm is being played simultaneously.

Ohm Crystal Bowls Activating an Ohm Crystal Bowl beneath a treatment table adds yet another dimension to a Sound Healing Treatment. With a yarn mallet, strike the side of the bowl periodically during the treatment. If your treatment table is made with wood, the cells of the wood will act as resonant chambers, harmonizing with the music, tuning forks and overall enhancing the treatment.

Contraindications

Of the treatments given, the following anatomical areas and points are contraindicated during pregnancy: Shoulder Well *(**Gall Bladder or GB 21**)* and the Low Abdomen *(**Sea of Qi – Ren 6**).*

Frequently Asked Questions

1. What part of the tuning fork is placed on the body?

Place the bottom of the stem of an activated tuning fork on the body. Be sure to apply the tuning fork with enough pressure to feel the vibration.

2. Is exact placement on an acupressure or acupuncture point necessary?

It is important to place the tuning fork as accurately as possible on the indicated point. Keep in mind that because sound vibration travels well through acu-points and meridians and through tissue, tendons, muscle and bone, exact placement of the tuning fork on a point is not required. However, since each point name carries with it a specific energetic and intention, accuracy is highly recommended.

3. What about contra-indications?

The same contraindications for acupuncture application apply to tuning fork placement during pregnancy: Shoulder Well (GB – 21) and the Low Abdomen (Sea of Chi – Ren 6).

4. Can the tuning forks be used on someone who has a pacemaker?

Yes, but if someone has a pacemaker it is contraindicated to place the tuning forks directly in the area of the heart. The Ohm Tuning Forks should be used distally and reflexively on other areas of the body.

5. How does the sound vibration of the tuning fork affect the person giving the treatment?

The Ohm tone from the tuning fork provides a consistent and protective field that helps shield the practitioner from any unwanted energy released during the treatment. Many of the same benefits extend to the person giving the treatment. For example, the sound vibration of Ohm helps keep the practioner grounded and clear as well as protected and nurtured by the healing qualities of this earth tone.

6. What is the Schumann Resonance? How does it differ from Ohm?

Schumann Resonance (SR) is sometimes referred to as earth tone. SR is based on the work of Winfried Otto Schumann (1888-1974, Germany), a scientist whose research is applied in the evaluation of the effect of climate change and global warming. Schumann resonances are global electromagnetic resonances, excited by lightning discharges in the cavity formed by the Earth's surface and the ionosphere. The lowest frequency mode of the SR occurs at a frequency of approximately 7.8 Hz. The ohm frequency also exists in this lower range, a lower octave being 8.5 Hz.

The following table highlights similarities and differences between SR and Ohm.

TABLE OF COMPARISON

	OHM	SR
Is this a scientifically measured Frequency?	yes	yes
What is the hertz measurement?	136.1hz*	7.83 - 60 hz
Is this measurement a constant?	yes	no
Is there a cyclical, rhythmic or seasonal nature to this tone?	yes	no
Is the measurement effected by human activity, e.g., electrical emissions & airborne pollution?	no	yes
Is this measurement effected by atmospheric activity, e.g., lightning storms, sun spots and other solar activity?	no	yes
Is there a spiritual, religious or philosophical basis for the determination of this frequency?	yes	no
Is this frequency used as the fundamental tone in contemporary sound healing systems?	yes	yes

*The Ohm frequency also exists in this lower range, a lower octave being 8.50 hz.

Sources

Amaro, Dr. John. *Electro Meridian Imaging.* International Academy of Medical Acupuncture, Inc., 2006.

Blavatsky, Helena. *The Secret Doctrine.* Theosophical University PR, 1999.

Carey, Donna and de Muynck, Marjorie. *Acutonics: There's No Place Like Ohm. Sound Healing, Oriental Medicine and the Cosmic Mysteries.* Devachan Press, 2002.

Cousto, Hans. *The Cosmic Octave.* Synthesis Verlag, 2004.

Deadman, Peter. *A Manual of Acupuncture.* Journal of Chinese Medicine Publications, 1998.

Gardner, Kay. *Sounding the Inner Landscape.* Element Books Limited, 1997.

Kaptchuk, Ted. OMD. *The Web That Has No Weaver.* Contemporary Books, 2000.

Kayser, Hans. Akroasis. *The Theory of World Harmonics.* Plowshare Pr, 1970.

Luce, Gay Gaer. *Biological Rhythms in Human and Animal Physiology.* Dover Publications, 1971.

Merryman, Marjorie. *The Music Theory Handbook.* Schirmer, 1997.

Narby, Jeremy. *The Cosmic Serpent. DNA and the Origins of Knowledge.* Penguin Putnam Inc., 1998.

Panchapakesan, Balaji. *http://www.udel.edu/PR/UDaily/2005/mar/nanobomb101305.html* Accessed June, 2008.

Parker, Barry. *Albert Einstein's Vision :Remarkable Discoveries That Shaped Modern Science.* Prometheus Books, 2004.

Pruden, Bonnie. *Pain Erasure. Discover the Wonders of Trigger Point Therapy.* M. Evans & Company, Inc., 1980.

Rael, Joseph. *Being and Vibration.* Council Oaks Books, 1993.

Rubin, Dr. Clinton. *http://science.nasa.gov/headlines/y2001/ast02nov_1.htm*. Accessed June, 2008.

Silva, Freddy. *Secrets of the Fields: The Science and Mysticism of Crop Circles.* Hamptom Roads Publishing Co., 2002.

Stone, Randolph. *Polarity Therapy.* CRCS Publications, 1986.

Tyme, L.Ac. *Student Manual on the Fundamentals of Traditional Oriental Medicine.* Living Earth Enterprises, 2001.

Upledger, John E. & Vredevoogd, Jon. *Cranial Sacral Therapy.* Eastland Press, 1983.

Vitebsky, P. *The Shaman.* New York: Little, Brown & Company. 2001.

Index

A
Absorption of sound by the body, 25, 26
Acoustic, 23
Activation of the tuning fork, 25–28
Activator, Practitioner, 25, 35, 43
Acu-points, 39, 42
Acupressure, 23, 39, 41, 77
Acupuncture
    accessing Qi through acupuncture points, 12
    Electro Meridian Imaging, 40
    history of, 41
    meridians, 40
    treatment protocols and, 39
    tuning fork as alternative to needles, 31
Acupuncture points (in treatment)
    GB 1 *(Fresh Innocent Eyes)*, 72
    GB 21 *(Shoulder Well)*, 69
    KI 1 *(Bubbling Spring)*, 64, 74
    LU 1 *(Central Treasury)*, 65
    REN 6 *(Sea of Qi)*, 50, 66, 68
    REN 17 *(Primordial Child)*, 49, 67, 68
    SI 11 *(Heavenly Gathering)*, 62
    ST 6 *(Mandible Wheel)*, 70
    ST 41 *(Stream Divide)*, 73, 74
    UB 2 *(Gathering Bamboo)*, 71
    UB 10 *(Heavenly Pillar)*, 58, 60
Acutonics, xv
Adrenals, 2, 3
Alapa (sitar pre-tuning), 19
Amaro, John, Dr., 39–40
Ankle pain, 73, 74
Anxiety, 50, 65, 68, 71
Application of tuning forks, 26–28
Application Process and Methods
    Mid, Low, and Osteo Ohm, 51
    High Ohm Octave, 53
    Sonic Ohm Octave, 53
Astronomy, 18–19
Aum. *See also* Ohm, Om, 15–16
Axis Mundi, 16, 55

B
Ba Liao, 57
Back. *See* Lower back; Spine
Balance
    boosting the immune system, 12
    High Frequency Tuning Forks restoring, 27–28, 35, 36
    natural cycles and, 6, 14, 17–18, 43
    Ohm Tuning Forks in meditation, 47
    response to sound in utero, 26
Bells, 19
Benefits of sound healing, 3
Bewitched, xv, 31
Bilateral Application, definition of, 51
Biorhythms, 6, 11, 13, 29, 37, 43
Blavatsky, Helena, 20
Blood flow, restoring, 57
Blurred vision, 71
Body attunement, 6
Body mechanics, 51–52
Body's healing intelligence, 21, 39
Bone density, increasing, xv–xvi, 34
Bowls, crystal, 76
Bowls, singing, 19, 22, 76
Breathing, deepening and enhancing, 48, 56, 61, 65, 67, 69, 70
Bubbling Spring (KI 1), 49, 64, 74

C
C (musical note), 16–17
Caduceus, 20
Caduceus Weaving Method, 53, 75
Calendars, octaves in, 21
Calming, 6, 48, 67, 76
Cancer, treatment of, xi, xv–xvi, 13, 14
Cancer cells
    characteristics of, xvi
    frequency applied to, 14
Cat purr, research data on, 34
Caulton, Donna, 7
Cayce, Edgar, 34
CDs, music, 37, 43
Centering, 66
Central Treasury (LU 1), 52, 65
Chakras, 28, 36, 57, 66, 67
Chemotherapy, 13
Circadian clock, 13, 17–18
Clenching, 70
Color in octaves and overtones, 21, 22
Complementary Alternative Medicine, 13
Concentration, 58
Connecting
    adult and child within, 68
    cerebral and physical, 60

to Earth, 49, 64
lower abdomen and heart, 50, 68
lower and upper body, 59
self and world, 48, 62
to universal energy (Qi), 58, 65
Contraindications for pregnancy, 48, 50, 66, 69, 76, 77
Cousto, Hans, 18
Cranial Sacral Therapy, 29
Crystal bowls, 76
Crystal clearing, 36
Cycles, natural
balance and, 6, 14, 17–18, 43
disconnection from, 6
electromagnetic frequencies disrupting, 11
importance of balance and homeostasis, 18
menstrual cycles, 6
octaves, 20–21
reconnecting through Ohm Therapeutics, xiii, xvi– xvii, 6
resonance with Ohm, 19–20, 37, 39
solar and lunar, 17, 18
use of sound in connecting with, 1
visual depiction of, 7
Cycles per second (cps), 11
Cymatics, 9–10

D
Depression, 58
Disease, definition of and about, xi, xvi, 1, 2, 12–14
Disharmony, 6, 23
Dissonance, 16
Distal Application, definition of, 51
DNA, 20
Double Application, definition of, 51

E
Earth
Earth energy, 36, 74
Earthly body, 3–4, 6
elliptical orbit, 14, 18
magnetic field, 11, 77
reconnecting to (*See* Cycles, natural)
tone of Om, 16–20, 36, 39, 43, 44, 77
Electro Meridian Imaging, 39–40
Electromagnetic frequencies (EMFs), 11–12
Electromagnetic radiation (EMR), 11
Eliot, T.S., 47
Energetic field, 35
Energetic pathways, 39, 42

Energy balance, 28, 36
Energy flow, treatment for, 59, 61, 62, 74, 75
Energy of the *Microcosmic Orbit*, 42
Entrainment, 5, 6, 17–18, 28, 39
Equal Tempered tuning, 17
Eyestrain, 71

F
Fear, 58
Fitzgerald, Ella, 13
Food, Sympathetic Resonance and, 5
Frequency
determining octaves, 20–21
Electro Meridian Imaging, 40
High Ohm Tuning Forks, 35
keynote of Ohm, 18
Low Ohm Tuning Fork, 34, 35
medical application of, 13–14
Mid Ohm Tuning Fork, 33
of Ohm, 17
Osteo Ohm Tuning Fork, 34
Sonic Ohm Tuning Fork, 35, 36
tuning fork length and, 24
Fresh Innocent Eyes (GB 1), 72
Fundamental tone, 17, 18, 19, 21, 22, 28, 39, 78

G
Gardner, Kay, 39
Gathering Bamboo (UB 2), 71
GB 1 *(Fresh Innocent Eyes),* 72
GB 21 *(Shoulder Well),* 48, 69, 76
Gladzewski, Andrew, 36
Grief, 65
Grounding, 49
Guiding Principles of Ohm Therapeutics, xxiii, 43

H
Harmonic Medicine, 9
Harmony, disruption of. *See* Disease
Harmony, restoration of, 12–14
Headaches, 48, 58, 60, 70, 71, 72, 73
Healing intelligence of body, 14, 21, 26, 28, 39
Healing properties of octaves, 21
Heart chakra, 67
Heavenly Gathering (SI 11), 62
Heavenly Pillar (UB 10), 58, 60
Heline, Corinne, 37
Hertz, 11
High Frequency Ohm Tuning Forks
Application Process and Methods, 27–28, 35–36, 53, 75
frequency of, 35–36

High Ohm Octave, 35, 36, 53
Hindu tradition, 10, 15
Holding sacred space, 45
Holistic treatment, 55
Homeostasis, xvii, xxiii, 12, 14, 18, 24
Hopi tradition, 10
Hua Tuo Jia Ji points, 52, 55, 57
Human conversation, 18
Human touch, 44
Huygens, Christian, 5

I
In the Key of Earth, CD, 12, 37, 43
Indigenous cultures, 1, 10, 19
Inflammation, 3, 34
Insomnia, 50, 68, 70
Intervals, musical, 17, 21, 22, 28, 34, 35
Intuition and intention, 4–5

J
Jaw, 70
Jenny, Hans, 9–10
Joints, benefits to, 3, 23, 34, 35
Joy, 74

K
Kepler, Johannes, 18
Key Points, 44
KI 1 *(Bubbling Spring)*, 64, 74
Kinetic energy, 10

L
Law of Octaves, 21
Lemniscate Music logo, 20
Listen, 56
Low frequency sound vibration, xv–xvi
Low Ohm Octave, 35
Low Ohm Tuning Fork, 34, 49–50
　Application Process and Methods, 26–27, 51
　frequency of and overtones, 22
Lower back, 59, 60, 63, 66, 74. *See also* Spine
LU 1 *(Central Treasury)*, 52, 65

M
Mandala, 10
Mandible Wheel (ST 6), 70
Mantra, 10
McClure, Jimmie, DC, 39–40
Medical application of frequency, 13
Medicine wheel, 45
Meditation, 47
Menstrual cycles, 6
Meridians, 39–40

Microcosmic Orbit, 41–42, 55, 57, 59, 66, 67, 75
Mid Ohm Tuning Fork
　Application Process and Methods, 26–27, 51
　frequency of, 18–19
Music
　crystal bowls, 76
　Mid Ohm frequency, 18–19
　octaves, 20–21
　Ohm and C, 16–17
　Ohm octave, 34
　Ohm vibration as musical center of gravity, 17–18
　overtones, 21–22
　for pain management, 37
　singing bowls, 76
　tuning forks and, 23–24, 37
　Western, 12
Music CDs, 37, 43
Music of the Spheres, 14
Music theory, 17
Music Therapy, 9, 13
Musical rest, 29

N
Natural cycles. *See* Cycles, natural
Nature, connection to, 18–20
Navajo tradition, 10
Neck pain, 69
Neck tension, 48, 58
Negative energy, 75
Nervous system, relaxing, 57
Newlands, John, 21
Nilsson, Lennart, 4
Noise pollution, 2–3, 11, 12
Novalis, 1

O
Octaves, 20–21, 34, 35
Ohm
　Ohm octave, 34
　Ohm singing and crystal bowls, 47, 76
　Ohm Therapeutics Guiding Principles, 23, 43
　and the rhythms of life, 39
　spellings of, 15–16
　symbol, 15–16
Ohm Tuning Forks
　acoustic advantage of, 23–24
　care of, 29
　history and diverse uses of, 31
　how to activate, 25
　how to hold, 24–25

length of vibration, 25
music and, 37
number of applications, 26
overtones, 22
parts of, 24
rhythm of application, 28
See also High Frequency; Low Ohm; Mid Ohm; Osteo Ohm; Sonic Ohm
Ohm vibration, 15–18
Om. See also Aum, Ohm 15–16
Orbit, Earth's, 14, 18
Orbit, Microcosmic. See Microcosmic Orbit
Osteo Ohm Tuning Fork
    Application Process and Methods, 26–27, 51
    frequency of, 34
Over-thinking, 71
Overtones, 21–22

P
Pace of application, 28–29
Pacemaker, 77
Paganism, 16–17
Pain relief, 34, 42
    ankle, foot, leg, 73, 74
    back, xv, 60, 74
    neck, shoulder, arm, 69
    shoulder, 62
Palpate, 44
Parasympathetic nervous system, 2
Para-vertebral points, 55
Periodic Chart of the Elements, 21
Piano keyboard, 16
Pitch, 11
Poetry, octaves in, 21
Pollution, 2–3, 11–12
Pregnancy, contraindications for, 48, 50, 66, 69, 76, 77
Primordial Child (REN 17), 49, 67
Psycho-spiritual illness, 6, 53
Pulse of the natural world, 37

Q
Qi (natural energy flow), 42, 44, 50, 57, 64, 65, 66

R
Rael, Joseph, 1
Rainbows, 22
Reflex points, 23, 41, 42, 55
Reflexology, 42–43

Relaxation
    chest and shoulders, 65
    entire body, 50, 56, 67, 68
    jaw area, 70
    legs and upper back, 63
    lower abdomen and lower back, 66
    through music, 37
    upper body, 48, 69
REN 6 *(Sea of Qi)*, 50, 66, 68
REN 17 *(Primordial Child)*, 49, 50, 67, 68
Resonance, 5, 19–20
Resonant energy, 75
Restlessness, 50, 68
Restoring energy and blood flow, 57
Rhomboid muscles, 61
Rhythm of application, 28
Rolling Method, 28, 53, 75
Rubin, Clinton, Dr., 34

S
Sacred space, 45
Sacrum, 59, 60
Sanskrit, 15–16
Schumann, Winfried Otto, 77
Schumann Resonance (SR), 77, 78
Sea of Qi (REN 6), 50, 66, 76
Self treatment, 47–50, 48–50
Shamanism, sound and healing in, 1, 2
Shore, John, 31
Shoulder pain, 62, 69
Shoulder tension, 48, 61, 62, 64, 65
Shoulder Well (GB 21), 48, 69, 76, 77
SI 11 *(Heavenly Gathering)*, 62
Silence, 29, 45
Singing bowls, 19, 22, 76
Single Application, definition of, 51
Sitar, 18–19
Solar year, 17
Sonar, 14
Sonic Ohm Octave, 35, 53
Sonic Ohm Tuning Fork
    Application Process and Methods, 27–28, 53
    frequency of, 35–36
Soothing and calming, 67
Sound
    electromagnetic frequencies, 11–12
    forms and characteristics of, 10–11
    healer, 29
    natural decay of, 29
    Ohm vibration, 15–16

resonating with Ohm, 19–20
waves, 11, 35
See also Vibration
Space clearing, 28, 36
Spine
   Hua Tuo Jia Ji points, 52, 55, 57
   Sacrum treatment, 59
   Spinal treatment, 57
   See also Lower back
ST 6 *(Mandible Wheel)*, 70
ST 41 *(Stream Divide)*, 73, 74
Stability and security, 64
Star Trek, 13
Still point, 29
Stomach, 70, 73
Stream Divide (ST 41), 73, 74
Strengthening, 63, 75
Stress relief, 65
   foot treatment, 64
   leg treatment, 63
   jaw treatment, 70
   rhomboid muscles, 61
   self treatment, 48
   shoulder well treatment, 69
Symbol, Ohm, 16
Sympathetic nervous system, 2
Sympathetic Resonance, 5–6, 44

T
Tension, releasing, 48, 49, 58, 61, 62, 64, 65, 69
Therapeutic applications
   acupuncture, trigger and reflex points, 41–42
   basic techniques, 43–44
   human touch, 44
   reflexology, 42–43
   self treatment, 47–50
   treating others, 50–53
   See also Treatment protocols
There's No Place Like Ohm, CDs, 37, 43
Threshold of noise, 12
Tibetan symbol, 16
Tonifying, 57
Tonoscope, 9–10
Treating others, 50–53
Treatment protocols
   abdomen, lower, 50, 66, 68
   ankle, 73, 74
   back, lower, 59, 60
   chest, upper, 49, 52, 65
   energetic field around body, 75
   eye region, 72
   eyebrow, 71
   foot, bottom of, 49, 64
   foot, top and bottom of, 74
   hand, 48
   head, back of, 58, 60
   heart chakra, 67
   Hua Tuo Jia Ji points, 52, 55, 57
   jaw, 70
   legs, back of, 63
   listening, 57
   pelvis, 59
   sacrum, 59, 60, 75
   shoulder, top of, 69
   shoulder blades, 61, 62
   spine, 55, 57, 59
   spine, cervical, 58, 60
   sternum, 49, 67, 68
   treatment space, 75
Trigger points, 42
Tumor, 14
Tuning forks. *See* Ohm Tuning Forks

U
UB 2 *(Gathering Bamboo)*, 71
UB 10 *(Heavenly Pillar)*, 58, 60
Unconditional love, 67
Unwelcome noise, 2, 3
Upanishad, Mandukya, 15
US Navy, 14

V
Vibration, 1–2, 9–11, 25. *See also* Ohm Tuning Forks; Sound
Vibrational Healing Music, CD, 37, 43
Vibrational Sound Healing, 13
Vibratory system, 3
Vision, 71, 72
Vitality, 66

W
Wand of Hermes, 53
Water in the earthly and physical body, 3–4
Wave theory, 9–10
Weights, tuning fork, 24, 29–30
West, John, 31
Western medicine, 13
Women, sympathetic resonance and, 6. *See also* Pregnancy, contraindications for
Wounded healer, xi

Y
Yantra, 10
Yoga, 37, 47

There's No Place Like Ohm™

About the Author

Born into a musical family, Marjorie de Muynck's interest in tuning forks, sound and vibration was awakened at an early age. Marjorie first held a tuning fork in 1963 to tune her guitar, and then began applying the sound vibration to her own body—a sign of things to come.

Marjorie's medicinal interest in Sound Healing and tuning forks grew and coalesced in 1995 when she co-founded the Kairo's Institute of Sound Healing, and co-developed the *Acutonics®* sound healing system and co-authored the textbook, *Acutonics: There's No Place Like Ohm, Sound Healing, Oriental Medicine and the Cosmic Mysteries*. Marjorie co-developed a healing treatment called *Harmonic Attunement®*, which incorporates the use of cosmically tuned symphonic gongs, tuning forks, didgeridoos, rattles, Tibetan bowls and bells. Marjorie also studied Huichol Shamanism with Brant Secunda for six years.

As a graduate of the ISIS School of Massage in 1983, Marjorie was a part of an early movement of professionals who helped to create certification standards for Massage Practitioners. As a prodigious student, practitioner and instructor of alternative healing modalities, Marjorie attended and graduated from the Seattle Shiatsu Institute (1983-1986) and has now studied Oriental Medicine for over 25 years. She taught Shiatsu, Hara Diagnosis, Kundalini Yoga, Reiki and Sound Healing at the Northwest Institute of Acupuncture and Oriental Medicine for 6 years.

In 1983 Marjorie co-created an On-Site Shiatsu business for King County and the City of Seattle, which she owned and operated for 17 years. Her treatments incorporated Sound Healing with tuning forks. During this period Marjorie also performed Native American flute on world renowned yoga teacher Ana forest's DVD *Strength and Spirit*.

Marjorie's eclectic mix of experience as a musician, composer, recording artist, healing arts practitioner and educator, combined with her scientific understanding of vibration and sound, led her to create *Ohm Therapeutics*™, a comprehensive vibrational healing system which combines the application of tuning forks with music composed in the healing frequency of Ohm.

Marjorie is a pioneer of musical compositions in the key of Ohm: *There's No Place Like Ohm Volumes 1 & 2* (Lemniscate Music, 2002 & 2005) and *In the Key of Earth* (Sounds True, 2007). At present, she is researching the acoustic properties of animal and plant sounds for her forthcoming CD, *Vibrational Healing Music* (Sounds True, March 2009). Each of these Sound Healing recordings are in Marjorie's signature tuning of Ohm.

Marjorie received her Master of Music in Music Education from Boston University, and her undergraduate degree in General Sciences with a minor in Music. Marjorie continues to further her medical studies at the Southwest Acupuncture College in Santa Fe.

Additionally, over the past 26 years her studies and certifications have included the following: Jin Shin Do, Kundalini Yoga, Iyengar and Hatha Yoga, Reiki, Jin Shin Jyutsu, Myofacial Release, Cranial Sacral Therapy, Hydrotherapy, Shen therapy, Polarity Therapy, Alexander Technique, Kinesiology, Ingham Method Reflexology, Chi Ne Tsang, Five-Element and Zen Shiatsu, Guided Imagery, Qigong, Pilates, NAET, Hara Diagnosis, Therapeutic Touch, Osteopathic Manipulation, Macrobiotic Diet, Ayurvedic Bodywork and Medicine.

Marjorie lives in rural Northern New Mexico and continues to find inspiration from her natural surroundings, the hummingbirds, crickets, and her friends and loving family.